TRANSPORT AND ROAD RESEARCH LABORATORY
Department of Transport

STATE-OF-THE-ART REVIEW 3

ROAD VEHICLE FUEL ECONOMY

by M H L

LONDON: HMSO

First published 1992
ISBN 0 11 551097 4

The views expressed in this review are not necessarily those of the Department of Transport

HMSO

HMSO publications are available from:

HMSO Publications Centre
(Mail, fax and telephone orders only)
PO Box 276, London, SW8 5DT
Telephone orders 071-873 9090
General enquiries 071-873 0011
(queuing system in operation for both numbers)
Fax orders 071-873 8200

HMSO Bookshops
49 High Holborn, London, WC1V 6HB 071-873 0011 (counter service only)
258 Broad Street, Birmingham, B1 2HE 021-643 3740
Southey House, 33 Wine Street, Bristol, BS1 2BQ (0272) 264306
9-21 Princess Street, Manchester, M60 8AS 061-834 7201
80 Chichester Street, Belfast, BT1 4JY (0232) 238451
71 Lothian Road, Edinburgh, EH3 9AZ 031-228 4181

HMSO's Accredited Agents
(see Yellow Pages)

and through good booksellers

Contents

Abstract

The large price increases of crude oil in 1974 and 1979 stimulated a search for improved fuel economy for road transport, but in the 1980's oil prices declined and interest in fuel saving decreased. Now the emission of carbon dioxide from road transport is accepted as one contribution to possible global warming, and the vulnerability of oil supplies to political events has been seen in the Middle East. Road vehicle fuel economy is again a topical subject.

This state of the art review from the Transport and Road Research Laboratory is largely based on Transport and Road Research Laboratory work over the past fifteen years. It examines the size and locations of oil reserves, and the demands likely to be made on them in the future by road transport. Principles for improving fuel economy by engine and vehicle design, and by the way motorists drive in real traffic are discussed. The potential of alternative fuels is covered, and the possible role of electric road vehicles assessed. The effects of particular tax regimes are briefly touched on.

All these subjects are illustrated by tables and diagrams, and the review contains over two hundred references to source material which will be useful to readers who want more detailed information.

1. Introduction

1.1 Background

The fuel economy of road vehicles is a subject whose apparent importance waxes and wanes with the price of petrol. In the early 1970's, it was only a few specialists who were well aware of limited reserves of natural crude oil, and of their concentration in unstable parts of the world. It was still possible, in 1970, for aviation economists to look forward to a continuing fall in oil prices for the next 20 years in looking at competition from trains and coaches (WPICT, 1970). As far as road transport was concerned, the area of greatest interest was the reduction of vehicle exhaust emissions for the improvement of air quality - driven in particular by the special conditions in the Los Angeles basin.

By the early 1980's, two major oil price rises, interruptions of supply from the Middle East, and widespread recognition that oil reserves were being used up rapidly, had all combined to provide considerable impetus to the search for more efficient use of fuel in road vehicles. Reduction of exhaust emissions was still important, but the realisation was growing that improved fuel economy and reduced emissions could be conflicting objectives. Research and development work on alternative fuels for road transport was in progress, with liquids derived from coal being a favoured possibility. Development of battery electric road vehicles was also active, mainly as a way of improving the air quality and reducing noise in cities, but also as a way of providing transport without dependence on oil. There were also hopes that electric urban buses and vans might prove commercially viable in a moderately generous tax situation (ie no vehicle excise duty and the absence of a tax on fuel).

In the 1990's, the situation has changed again. The late 1980's saw widespread concern about possible long term global warming brought about by the increasing emission into the atmosphere of the so-called "greenhouse gases". While there are many gases which contribute to the greenhouse effect, one of the major contributors is carbon dioxide. This is an almost inevitable result of burning fossil fuels like coal and oil, and the main way of reducing the rate of emission is to burn less fuel, either by using less of the product (eg electricity, or transport) or by using the fossil fuel more efficiently. The concern over carbon dioxide emissions has been a new stimulus to the search for greater fuel economy in road vehicles. In the UK, for example, road transport contributes about 18% of CO_2 emissions, and accounts for perhaps 10%-15% of UK's total effect on global warming. Road transport is not the major culprit, but it is an important source of CO_2 emission which is likely to grow as traffic increases over the next 25 years. Clearly there is a renewed interest in measures for reducing CO_2 emissions from road transport, and thus renewed emphasis on fuel economy.

The global warming/enhanced greenhouse effect mechanism is still the subject of scientific controversy. This review does not pretend to make any definitive statements on the subject, but a summary of the basic atmospheric physics, and the role of road transport, is given in Appendix A.

The other major change in the 1990's is the greater emphasis which is likely to be given in the future to security of supply of natural oil, following from the 1990 Gulf Crisis, and the 1991 Gulf War, which has emphasised the vulnerability of oil supplies. The knowledge that over 60% of proved oil reserves are in the Middle East, and that (at current production rates) they will last for more than 100 years, makes a return to peace and stability in this area of the greatest importance to oil-importing countries.

For a number of old, and new reasons, the interest in road vehicle fuel economy, and ways to improve it, is increasing again, after a period in the mid 1980's when falling oil prices made the subject unfashionable.

1.2 Purpose and scope of the review.

The main purpose of this review is to bring together, and to up-date where necessary, the research and assessment work on vehicle fuel economy, alternative fuels, and electric vehicles which was carried out as part of the Transport and Road Research Laboratory's programme from the early 1970's up to the present day. While some of the work is ten years old, and more, it has been encouraging to find that many of the studies and much of the experimental work have relevance to the problem of the 1990's, and can still provide useful guidance.

The scope of the review is not confined to TRRL work, nor with the situation in the UK. The published work from other organisations, and from international bodies like the OECD and the International Energy Agency are drawn on as necessary.

The review has been written with the intention of interesting a reader with a general technical and scientific background, but no particular expertise in fuels, vehicle and engine design, or road construction. The reader will not find many mathematical formulae, but will be provided with Tables and Figures to illustrate and explain the main features of the subject. An important objective is to provide, in one document, comprehensive references to the TRRL work, and to the work of others, which has up to now been scattered about in many publications. Because of the number of references, they are collected together for convenience at the end of the Review. While this leads to some duplication, the Chapters are largely self-contained, and it is hoped that the reader will excuse the duplication as a minor drawback to easier location of the papers, reports and books which have been used.

1.3 Topics included

The increased attention now being given to the emission of greenhouse gases by road transport is covered in Appendix A. Much of the information on the physics of global warming comes from a valuable report by the Select Committee on Science and Technology (House of Lords, 1989,

I and II). The Energy Committee (House of Commons, 1989, I, II and III) made useful comments on policy aspects of the greenhouse effect, and related transport matters. The White Paper "This Common Inheritance" (1990) has also been used. References to carbon dioxide emissions and to these sources of information occur throughout the review, as well as in the Appendix.

The body of the review starts, in Chapter 2, with an examination of world fuel consumption, the importance of natural oil, and the demand for petroleum products by road transport in the UK. Projections of future use are then considered in the light of known oil reserves. The Chapter ends with a brief look at possible alternative fuels.

In Chapter 3, the effects of technical factors on road vehicle fuel economy are described. Particular topics treated are the comparison of petrol and diesel cars and vans, the effects of short journeys with cold engines, transmission systems, and the (often deleterious) effects of more stringent exhaust emission controls. The Chapter ends with some discussion on future possibilities for the improvement of fuel economy.

The review then moves (in Chapter 4) beyond the vehicle to topics relating to the driver and traffic and their influence on fuel economy achieved in practice. Some of the suggestions for economy aids and for driver training are assessed.

The next Chapter examines the extent to which taxation can encourage fuel efficiency. It sketches the connection (historical, and present) between the provision and maintenance of the road system, and the taxes paid by motor vehicles. The UK system of taxation on vehicle purchase, ownership and use is compared with other countries in order to illustrate the possibilities of using a tax regime to reduce total fuel consumption. The prevalence of "company cars" in the UK is considered, together with the likely effects of changing the system of paying for car business travel with the aim of encouraging greater emphasis on fuel economy.

Chapter 6 returns to the possible alternatives to natural oil for free-moving road transport, and considers the pros and cons of liquids from coal, methanol, and hydrogen. The potential of battery powered vehicles as an alternative to the direct use of oil-based fuel is discussed in Chapter 7, considering both the short to medium term developments and the long term prospects.

The review concludes (Chapter 8) with a summing up of the main lessons drawn from the earlier Chapters. One new topic is introduced, the comparison of the energy use of road transport and other ground modes (like trains), and in particular the public transport versus private transport competition for passenger travel. The details of this topic are given in Appendix B, and only the main conclusions are summarised in Chapter 8. The source of much of the information in Appendix B comes from a very thorough report by Martin and Shock (1989), which has also provided new information for other Chapters.

As will have become clear, this review has concentrated on land transport, by road except for the inter-modal comparisons in Chapter 8 and Appendix B. Air transport is not included. This is not because it is a trivial user of fuel. It is a fast growing sector, growing as quickly as road transport over the past decade, and in the European Community it used in 1986 about 10% of total transport fuel (CEC, 1988). It is also important because it competes for the "fraction" of crude oil which is used by diesel powered road transport, which (as will be seen) is particularly efficient. For these reasons, air transport should not be ignored as an indirect influence in the fuel economy of road transport, but it is a large subject and outside the scope of the present review.

The main purpose of the review is to give the concerned reader an overview of the "state of the art" in road vehicle fuel economy, and to identify the main source references for the reader who wishes to dig more deeply.

2. The use of oil-based fuel: Past, present, and future

2.1 World fuel consumption

The continuing importance of fuel derived from crude oil can be seen by looking at statistics for world consumption of primary energy between the years 1970 - before the first "oil crisis" - and the late 1980's. The published data has at least one shortcoming. It does not include the use as fuel of wood, peat or animal waste, though these are not negligible sources of fuel for many of the less developed countries. Unfortunately, the use of these fuels is unreliably documented, so that in the discussion which follows they are left out of the account. One estimate is that the use of wood alone might have increased the world primary energy consumption, as statistically recorded, by about 15% in the mid-1970's (Foley, 1976 and Openshaw, 1980).

With this omission in mind, the published statistics (BP, 1989) give the world consumption of primary energy in 1988 as about 8000 million tonnes of oil equivalent (MTOE). About half the world total is used by the developed nations who are members of OECD*.

The 1988 world total was over 50% greater than the total consumption in 1970. Nearly half (45%) of the increase has taken place in the Socialist Countries as defined in Table 2.1. About one quarter (27%) has taken place in OECD and the same proportion has taken place in the rest of the Non Communist World. Table 2.1 gives the detailed figures.

Within the total primary energy consumption, oil provided nearly 40% of the fuel in 1988 for the whole world, nearly 45% for the Non Communist World, and only slightly less (43%) for the OECD countries (Table 2.2 gives the details). The corresponding figure for the main OECD countries in 1970 was nearly 53% (BP,1971) indicating how the price changes and uncertainties of supply since 1970 have tended to make other fuels preferred to oil. Never-the-less, oil remains the most important single source of primary energy for the world as a whole, and certainly for the developed countries. The way consumption of different fuels has changed over the period 1970 to 1989 is illustrated in Figure 2.1 for the whole world. It can be seen there that most of the growth has been in the use of non-oil sources of energy, particularly coal and nuclear.

2.2 Fuel use in the United Kingdom

The situation in the United Kingdom in 1988 is very similar to the world consumption proportions of the different fuels, (rather than the OECD). The UK's use of coal, for instance, is

*See Table 2.1 for definitions.

TABLE 2.1
World energy consumption by area: 1970 AND 1978

(MTOE)	*1970*	*1988*	*Increase 1970-88*
OECD Countries[1]	3232.4 (100%)	4010.5 (124%)	778 (27%)
Rest of non-communist world	515.3 (100%)	1317.8 (256%)	803 (27%)
Socialist countries[2]	1425.1 (100%)	2745.2 (193%)	1320 (46%)
Total world consumption *	5172.8 (100%)	8073.5 (156%)	2901 (100%)

* Excluding wood, peat and animal waste used as fuel.

Source: BP, (1989)

Notes

(1) The original members of OECD are:-
Austria, Belgium, Canada, Denmark, France, West Germany, Greece, Iceland , Ireland, Italy, Luxembourg, the Netherlands, Norway, Portugal, Spain, Sweden, Switzerland, Turkey, the United Kingdom, the United States. Later signatories are Japan, Finland, Australia and New Zealand.

(2) Socialist countries are defined as:-
China; USSR; Albania; Bulgaria; Cuba; Czechoslovakia; East Germany; Hungary; Kampuchea; Laos; Mongolia; North Korea; Poland; Romania; Vietnam; Yugoslavia.

The country groupings are made purely for statistical purposes, and are not intended to imply any judgement about their political or economic standing. This is particularly so since the major changes in Eastern Europe in 1990.

TABLE 2.2
World energy consumption by fuel: 1988

(MTOE)	*Oil*	*Nat gas*	*Coal*	*Nuclear/Hydro*	*Total*
World total	3039	1631	2428	976	8074
consumption	(37.6%)	(20.2%)	(30.1%)	(12.1%)	(100%)
Of which:					
OECD	1715	763	899	634	4011
	(42.8%)	(19.0%)	(22.4%)	(15.8%)	(100%)
NCW	2374	967	1181	806	5328
	(44.6%)	(18.1%)	(22.2%)	(15.1%)	(100%)

Source: BP, (1989)

TABLE 2.3
World and UK energy consumption by fuel: 1988 (and 1970)

(MTOE)	*Oil*	*Nat gas*	*Coal*	*Nuclear/Hydro*	*Total*
Total world 1988	3039	1631	2428	976	8074
	(37.6%)	(20.2%)	(30.1%)	(12.1%)	(100%)
UK consumption 1988	78.3	47.9	65.9	17.9	210.0
	(37.3%)	(22.8%)	(31.4%)	(8.5%)	(100%)
1970	98.2	10.5	92.3	7.0	208.0
	(47.2%)	(5.0%)	(44.4%)	(3.4%)	(100%)

Sources: BP, (1989); Dept of Energy (1989a); TRRL (1981)

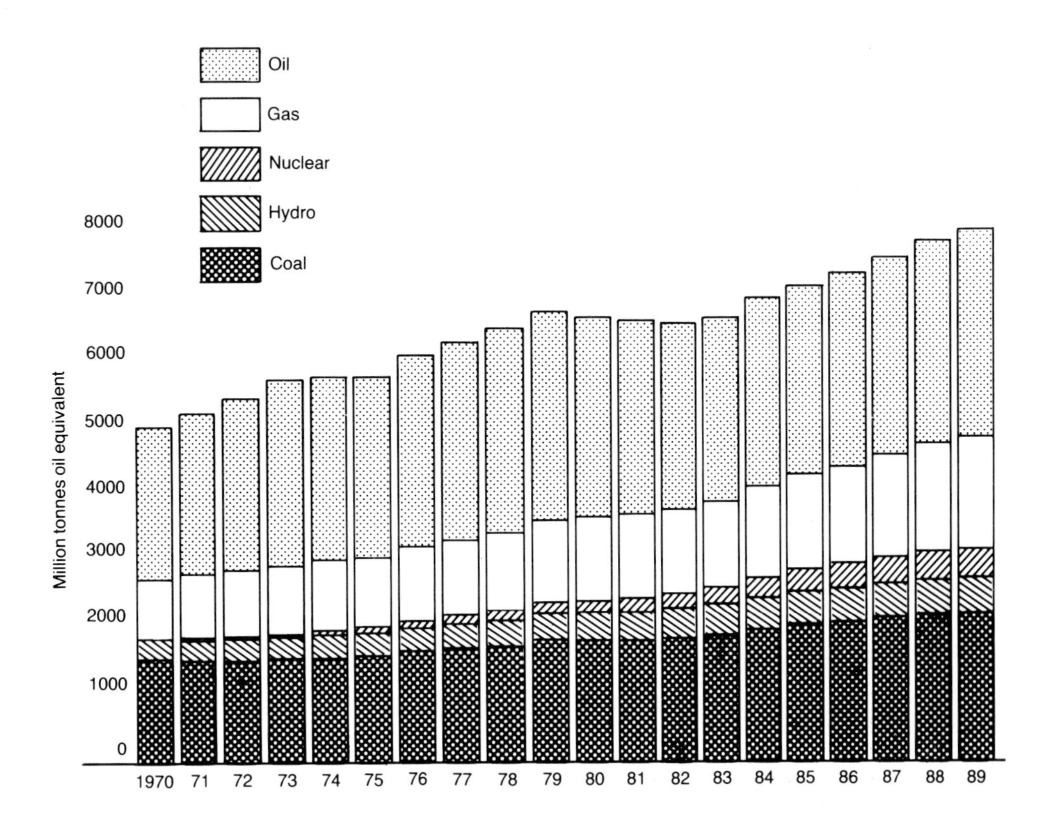

Fig 2.1 *World consumption of primary energy: 1970-1989 (Source: BP, 1990)*

higher than in OECD as a whole, and the proportion of total energy supplied by oil consequently rather less. Table 2.3 shows the comparative figures, and gives the UK values for 1988 and for 1970. The most notable change between the two years, where absolute energy consumption has hardly changed, is the reduction in consumption of oil and coal, and their replacement mainly by natural gas, and to a lesser extent by nuclear power.

Detailed comparisons for six years between 1970 and 1988 are given in Table 2.4, and the main features shown in Figure 2.2. The use of petroleum peaked in 1973, before the first major price rise, when it accounted for about 50% of UK's energy consumption. It then fell from the peak value of 108 million tonnes to around 70 million tonnes in 1983 before beginning to rise again. (The sharp increase in 1984 was caused by more oil being used for electricity generation during the long miners' strike.) By 1988, the UK was using 78 million tonnes, and this was only 37% of primary energy. The various end uses for which petroleum products are supplied are set out in Table 2.5 for the period from 1970 to 1988, and the main characteristics for the two years 1970 and 1988 are shown diagrammatically in Figure 2.3. Total primary energy consumption is nearly the same in both years (208 MTOE and 1970, and 210 MTOE in 1988), but while the petroleum used is 10 percentage points lower in 1988, the percentage of petroleum used for transport has

TABLE 2.4
The consumption of primary fuels and equivalents in the UK
Million tonnes of oil or oil equivalent (%)

Year	*1970*	*1973*	*1977*	*1980*	*1985*	*1988*
Coal	92.3 (44.4)	78.2 (35.6)	72.2 (34.6)	71.1 (35.6)	62.0 (30.9)	65.9 (31.4)
Petroleum (inc. non-energy use)	98.2 (47.2)	108.2 (49.3)	90.0 (43.1)	78.3 (39.2)	76.2 (37.9)	78.3 (37.3)
Natural gas	10.5 (5.0)	26.0 (11.8)	36.9 (17.7)	41.4 (20.7)	48.4 (24.1)	47.9 (22.8)
Primary Electricity (nuclear and hydro)	7.0 (3.4)	7.1 (3.2)	9.6 (4.6)	9.0 (4.5)	14.3 (7.1)	17.9 (8.5)
Gross inland consumption	208.0 (100.0)	219.5 (100.0)	208.7(100.0)	199.8 (100.0)	200.8 (100)	210.1 (100)

Sources: Dept of Energy (1989a); TRRL (1981)

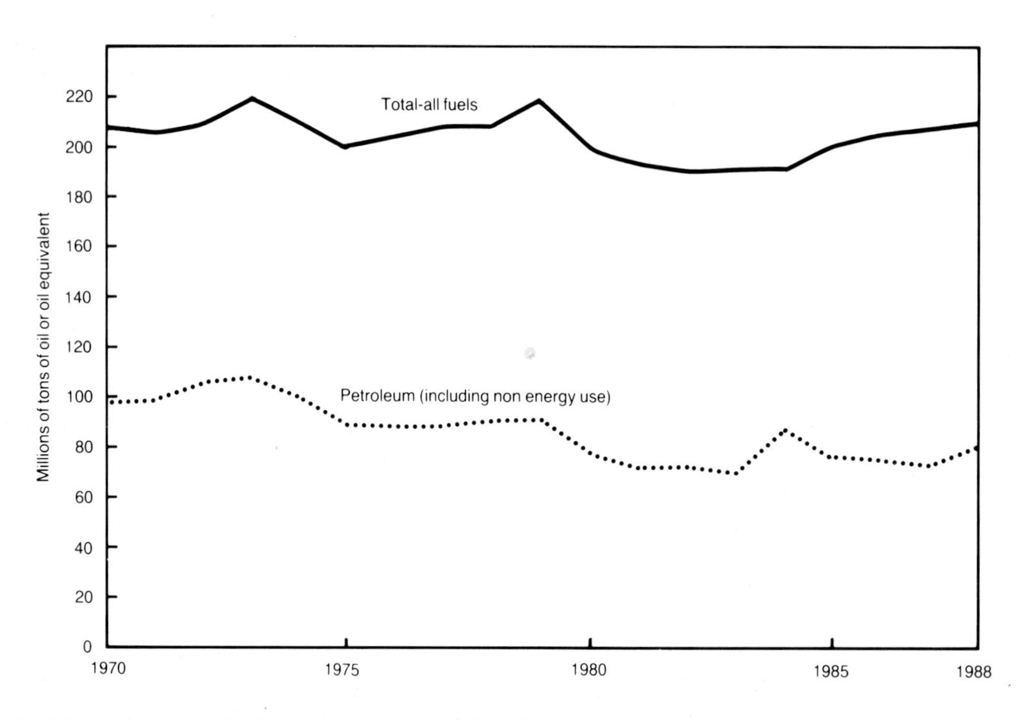

Fig 2.2 Gross UK inland consumption of primary fuels (1970 to 1988)

TABLE 2.5
The consumption of refined petroleum products in the UK: 1970 TO 1988
Millions of tonnes (%)

Year		*1970*	*1973*	*1980*	*1988*
Products used for energy					
Consumption by fuel producers		17.2 (17.5)	19.3 (17.8)	6.8 (8.7)	4.5 (5.7)
Industry		27.2 (27.6)	27.2 (25.1)	16.2 (20.7)	9.6 (12.3)
Road Transport		19.3 (19.6)	22.6 (20.9)	25.0 (31.9)	32.6 (41.7)
Other Transport	Air	3.5	4.3	4.7	6.2
	Rail	1.2	1.0	0.9	0.7
	Water	1.1	1.0	1.2	1.1
Other Transport Sub-total		5.8 (5.9)	6.3 (5.8)	6.8 (8. 7)	8.0 (10.2)
Other energy use		11.6 (11.8)	12.8 (11.8)	9.4 (12.0)	6.9 (8.8)
Total products used for energy		81.0 (82.3)	88.2 (81.5)	64.2 (81.9)	61.6 (78.6)
Non energy use		10.1 (10.3)	11.6 (10.7)	7.0 (8.9)	10.0 (12.8)
Total deliveries		91.2 (92.7)	99.8 (92.2)	71.2 (90.8)	71.6 (91.4)
Use in refineries and losses		7.2 (7.3)	8.4 (7.8)	7.2 (9.2)	6.7 (8.6)
Total consumption		98.4(100.0)	108.2(100.0)	78.4(100.0)	78.3(100.0)

Series: TRRL (1981); Dept of Energy (1989a)

almost doubled, and for road transport has increased from 20% in 1970 to 42% in 1988. The tonnage of petroleum used for road transport over this period has obviously increased. Figure 2.4 shows that the total has risen from nearly 20M tonnes to over 32M tonnes. Consumption of motor spirit (for cars, light goods vehicles and motor cycles) has increased quite steadily over the period, only slightly affected by the major price rises of crude oil in 1973 and 1979. The quantity of diesel fuel (for heavy goods vehicles, buses and most taxis) has also increased, from about 5M tonnes to over 9M tonnes, with most of the increase concentrated in the last part of the 1980's when the UK economy entered a period of growth.

2.3 Oil prices and elasticity of demand

The figures reviewed above show the expected reaction of consumers, other than road transport, to the price rises imposed by OPEC, the oil producer's cartel, in the early and late 1970's. The scale of the price increase is shown in Figure 2.5, and the modifying effect of UK inflation is illustrated in the upper dotted line. While the UK Retail Price Index is not necessarily the best indicator, it is sufficient to show that the "real" price of crude oil peaked in 1974 and reached a maximum in 1981 at over 70 US $ per barrel (in 1991 money), and since 1981 has fallen substantially to a level not far off the pre-1973 price. The same picture emerges on a longer time-scale (back to 1900) in Figure 2.6. Even the 1990 Gulf crisis, which pushed oil prices briefly up to 38 US$ per barrel, has not made crude oil exceed the 1981 "real" price.

The rise in price in the mid 70's and early 80's has had the result of concentrating consumption into premium applications, and away from applications where alternative fuels are available. For the UK, as Table 2.5 has shown, oil consumption (as a percentage of total use) in the period 1970

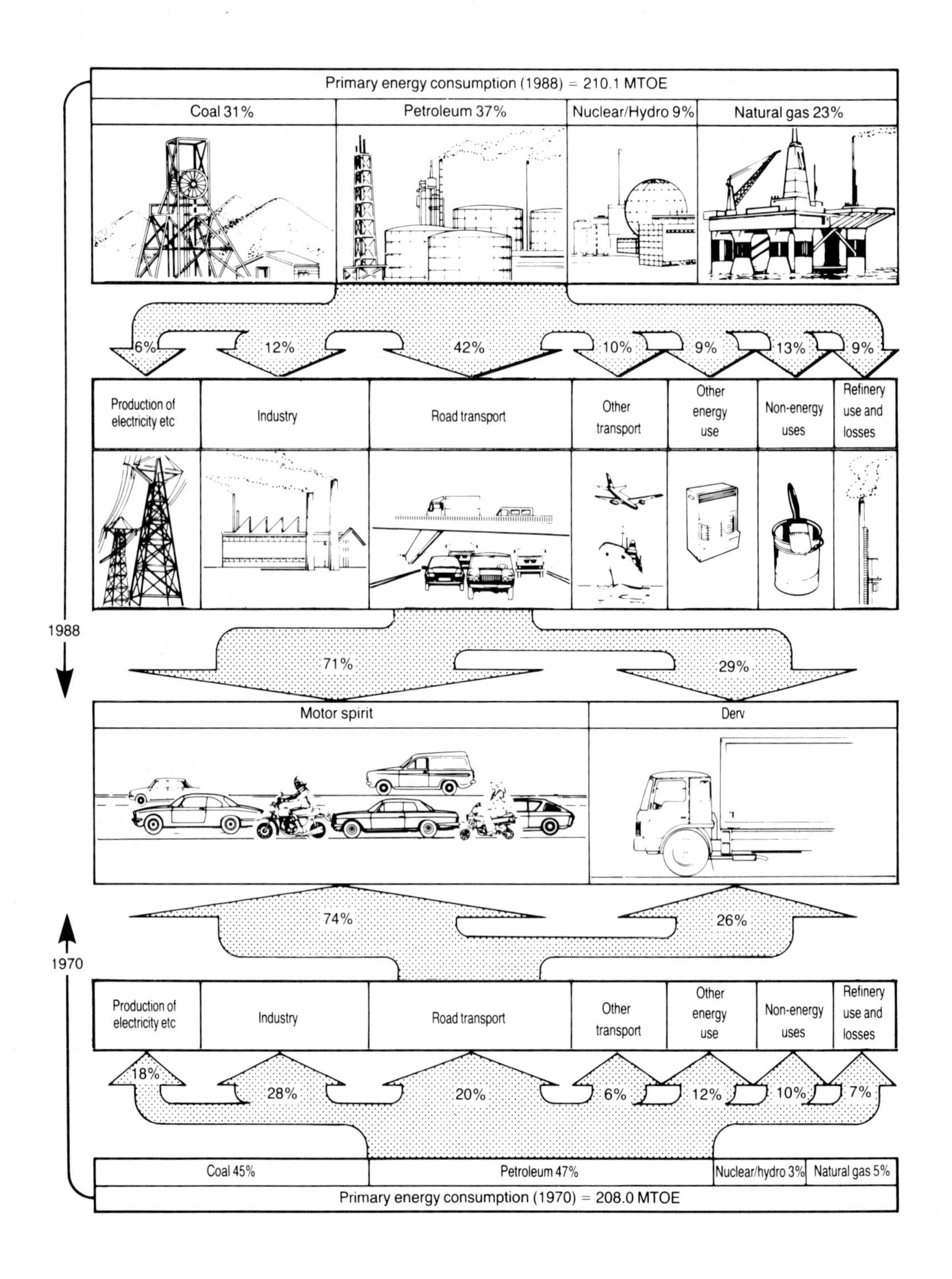

Fig 2.3 Energy consumption in the UK: 1988 and 1970

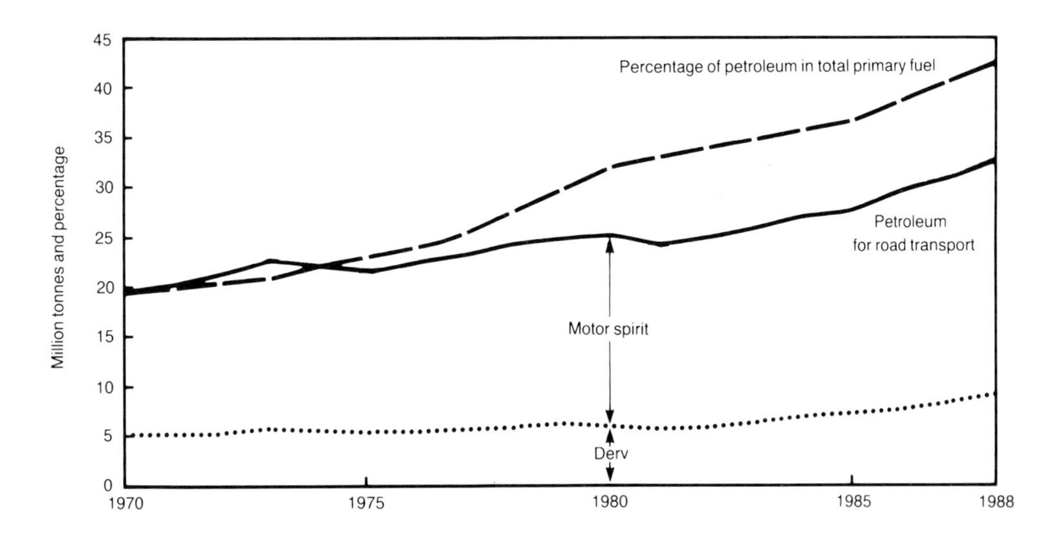

Fig 2.4 UK deliveries of fuel for road transport (1970 to 1988)

to 1988 has increased in the transport sector, but reduced considerably in industry. The reduction in industry's share is partly due to the low growth in the UK economy up to the mid 1980's, but there is little doubt that the use of alternative fuels like natural gas and electricity in preference to oil has been a major factor.

In the transport sector, and particularly for road transport, no alternative fuel is readily available, though rail electrification introduces a flexibility of fuel for power generation which is denied to diesel locomotive haulage. But rail uses only about 3% of total transport energy in the UK, so the effect is small, and the dominant factor in transport is the need for a portable liquid fuel for free-ranging road transport.

One reason why demand for petrol products for road transport in the UK has been relatively insensitive to the price of crude oil is that tax, in the form of duty and VAT, is levied at a high and time varying rate on fuel used for road transport, and the high inflation in the mid 70's reduced real price rises. Figure 2.7 shows the changes in price of motor spirit, both in terms of "money of the day" and in "1988 pence" using the Retail Price Index. It is clear from the chain dotted curve that the large increases in oil prices have been considerably damped by the large component of taxation in the total price, and by general price inflation.

The effect of price changes in demand for road transport fuel (elasticity of demand) has been studied by many authors for different purposes (for example, Bland (1984) as an input to transport modelling, and Tanner (1981) as an input to national travel forecasts). The results are far from straight forward. If attention is restricted here to the elasticity of average annual car use (yearly car km) with petrol price, Oldfield (1980) obtains values in the range -0.10 to -0.17 for Great

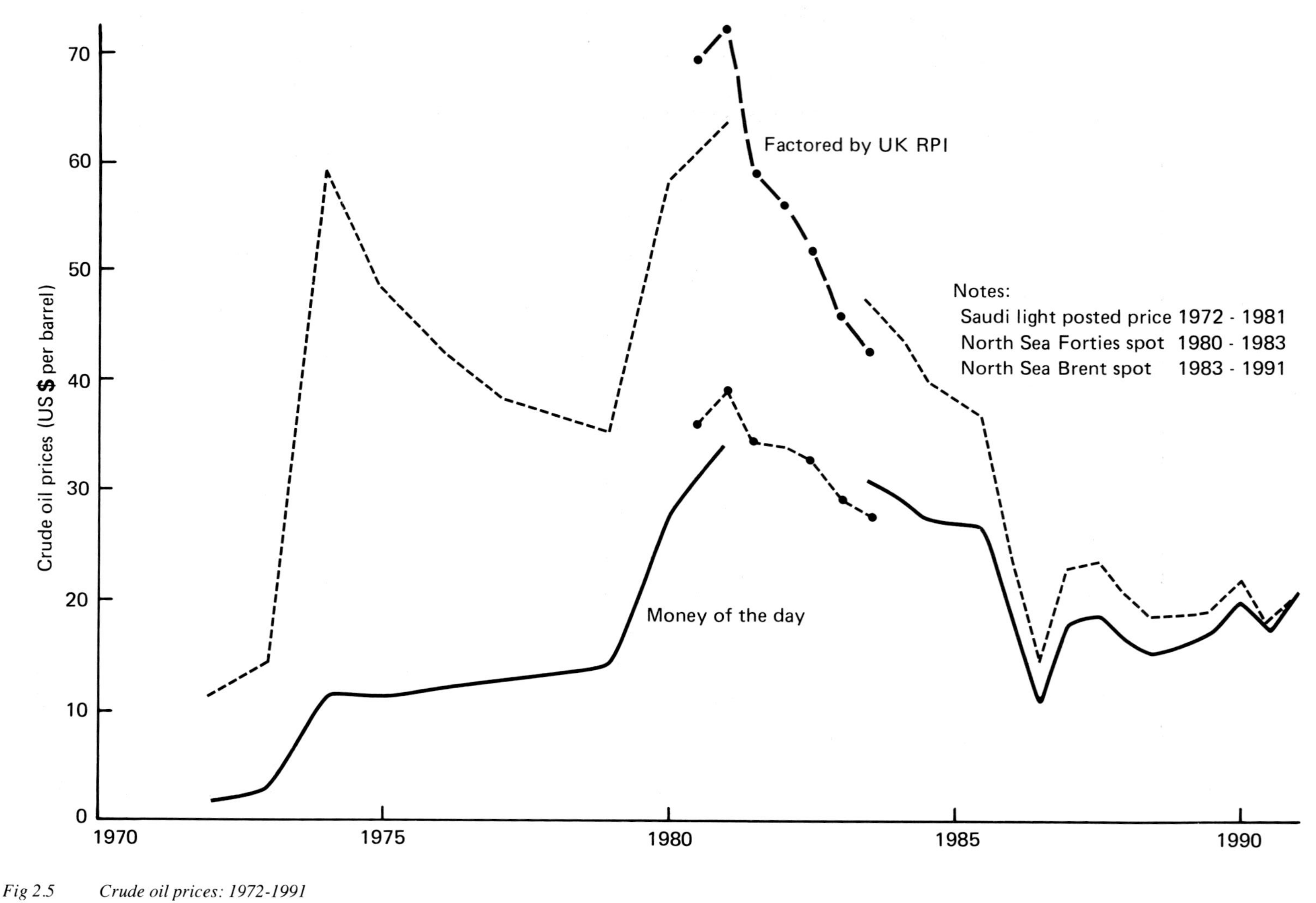

Fig 2.5 *Crude oil prices: 1972-1991*
(Source: Institute of Petroleum, 1989, with updates)

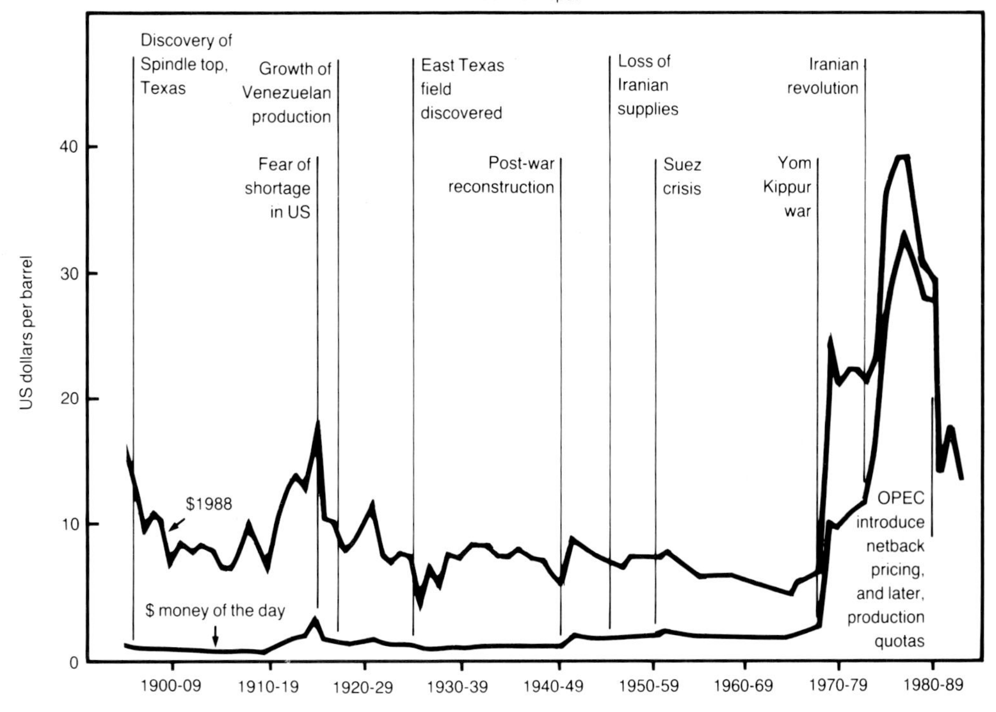

Fig 2.6 *Crude oil prices since 1990 (Source: BP, 1989)*

Britain in the period 1972-78.* Tanner (1981) studied the period 1960-1979 and derived an estimate of -0.15, but suggested that a value as low as -0.10 could not be ruled out. In his model for car use, an elasticity with petrol price of -0.12 was used.

The results quoted were obtain from time-series analysis, and would be expected to be applied to short term trends. In the longer term, the car user may well react to higher petrol prices by buying more efficient or smaller cars and reducing the amount of travel as well. In an international comparison, Tanner (1983) found that the quantity of fuel used per car across 19 countries gave an elasticity with price of just under -1.0: that is, if petrol cost 50% more, then just over 2/3 as much would be used. This effect appeared to operate mainly through reduction in the size of car.

---

*An elasticity of -0.10 means that, if petrol price increases by 1%, annual car kilometres are reduced by 0.1% approximately. The relationship is:

car kilometres = $a(P)^e$

where: a = a constant
P = petrol price
e = elasticity

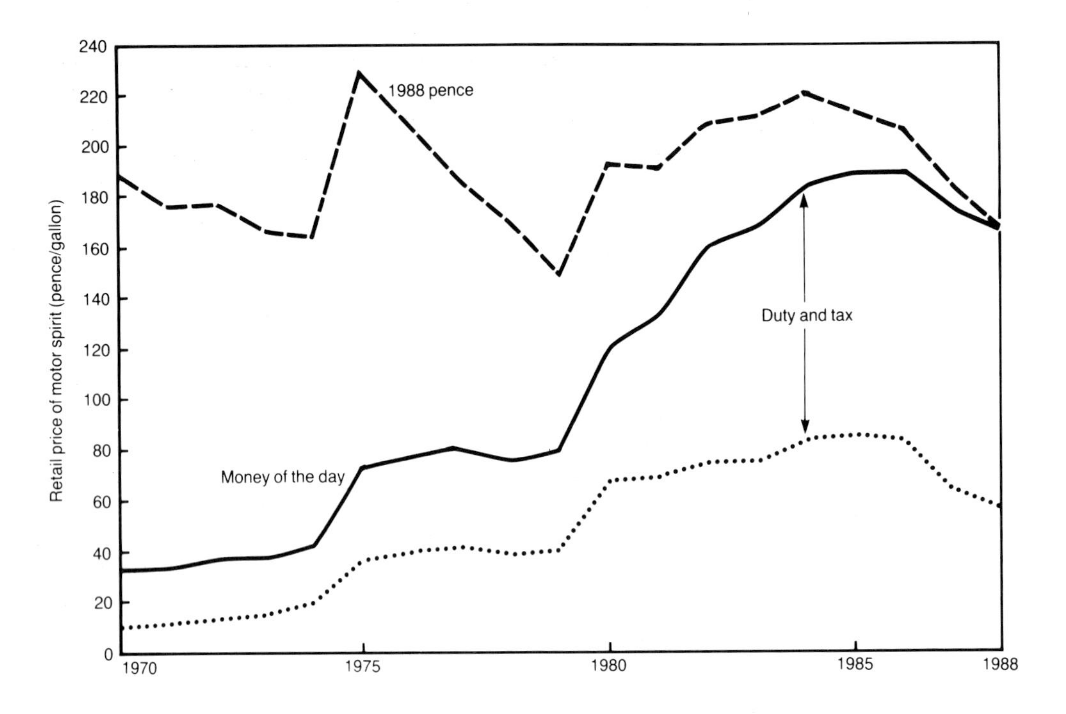

Fig 2.7 Retail Price of Motor Spirit 1970-1988

Analysis of short term fluctuations gave an average over the countries of -0.27, close to an average value for elasticities of kilometres per car which was found to be -0.25.x

Within these average figures there was some difference between countries. The UK was lower than average - for short term fluctuations the value was -0.15 compared with the average of -0.25.

Another international study by the IEA (1984) covered the experience of different countries over the period 1971-1981, and the results are given in Table 2.6 for the average distance travelled by

TABLE 2.6
The effect on average distance travelled by passenger cars of
(i) Growing/Affluence (ii) Petrol prices (iii) Growing car ownership

		Historical Elasticities with respect to:		
	Estimation Period	(i) GDP	(ii) Real Petrol Price	(iii) Car Ownership
United States	1971-81	0.475	-0.171	-0.575
Canada	1971-81	0.556	-0.181	-0.632
Japan	1973-81	0.535	-0.184	-0.506
Germany	1973-81	0.501	-0.152	-0.480
United Kingdom	1971-81	0.266	-0.091	-0.197
Italy	1973-81	0.735	-0.286	-0.601

Source: IEA (1984)

passenger cars as influenced by (short term) historical elasticities with respect to Gross Domestic Product, real petrol price, and car ownership.

It is noticeable again that the elasticities for the UK are markedly lower for all factors than those of the other countries. The elasticity for petrol price at about -0.10 is consistent with the results of Oldfield (1980) and Tanner (1981). The IEA suggest that one reason for the low values in the UK may be the high proportion of company cars (around 45% of new car sales in 1984) which would make car use by these car owners relatively immune from income and price changes. The predictive accuracy of the elasticity models is illustrated in Figure 2.8, which for six different countries shows reasonable agreement between observed car use and the model results.

More recently, Rice and Frater (1989) have investigated the demand for petrol in the UK over the period 1977-86 and included an explicit term for the potential fuel consumption rate of cars by their age in the fleet. The fuel consumption figures which are used are the detailed results for each model type published by the Department of Transport (see Chapter 4). It is shown that using a new car consumption term allows a better representation of the structure and mean speed of response of demand to price effects. The "new model year" car fuel elasticity, as defined in the paper, was found to be about +0.10, in the sense that a 10% decrease in litres/100 km (10% increase in efficiency) over the previous model year decreased total petrol demand by 1%. The authors claim that if the term is left out, the calculated short-term price elasticity is reduced, which could be misleading. For further details of a complex subject, the reader should refer to the original papers which have been cited.

2.4 Road traffic and vehicle fuel consumption

The result on road traffic - the measure of vehicle use - of the conflicting factors of oil price, general prosperity and average vehicle fuel efficiency is shown in Table 2.7 and illustrated in Figure 2.9 for the OECD countries and for Great Britain. Over the period 1970 to 1985, total road traffic (in million vehicle kilometres) has increased by 62% for OECD (54% for G.B.), while for passenger cars, the increase is 55% for the OECD countries (59% for GB*). Within the OECD countries, there are considerable variations, from only 42% for North America to nearly 130% for Japan. However, the experience of traffic growth in Great Britain is broadly similar to the OECD average.

Changes in overall road vehicle fuel consumption between 1970 and 1988 can be examined in an approximate way on the basis of fuel delivered and traffic volumes for Great Britain. It is possible to look at trends in petrol driven vehicles, and those fuelled by derv separately, though, as Rice (1982) also found, there are difficulties in reconciling the type of fuel used with the definitions of vehicle class used in the traffic counts. As a first step, travel by cars, motor cycles and light vans is correlated with petrol used, and travel by heavy goods vehicles (HGV's), large buses and coaches with derv consumption. There are obvious errors introduced by diesel powered cars, taxis, and light vans whose mileage is included in the "petrol user" classification, and these difficulties are addressed later. The basic data (and sources) are given in Table 2.8.

---

*Though the GB figures include travel by light goods vehicles (vans) which the OECD figures put into the goods vehicle total.

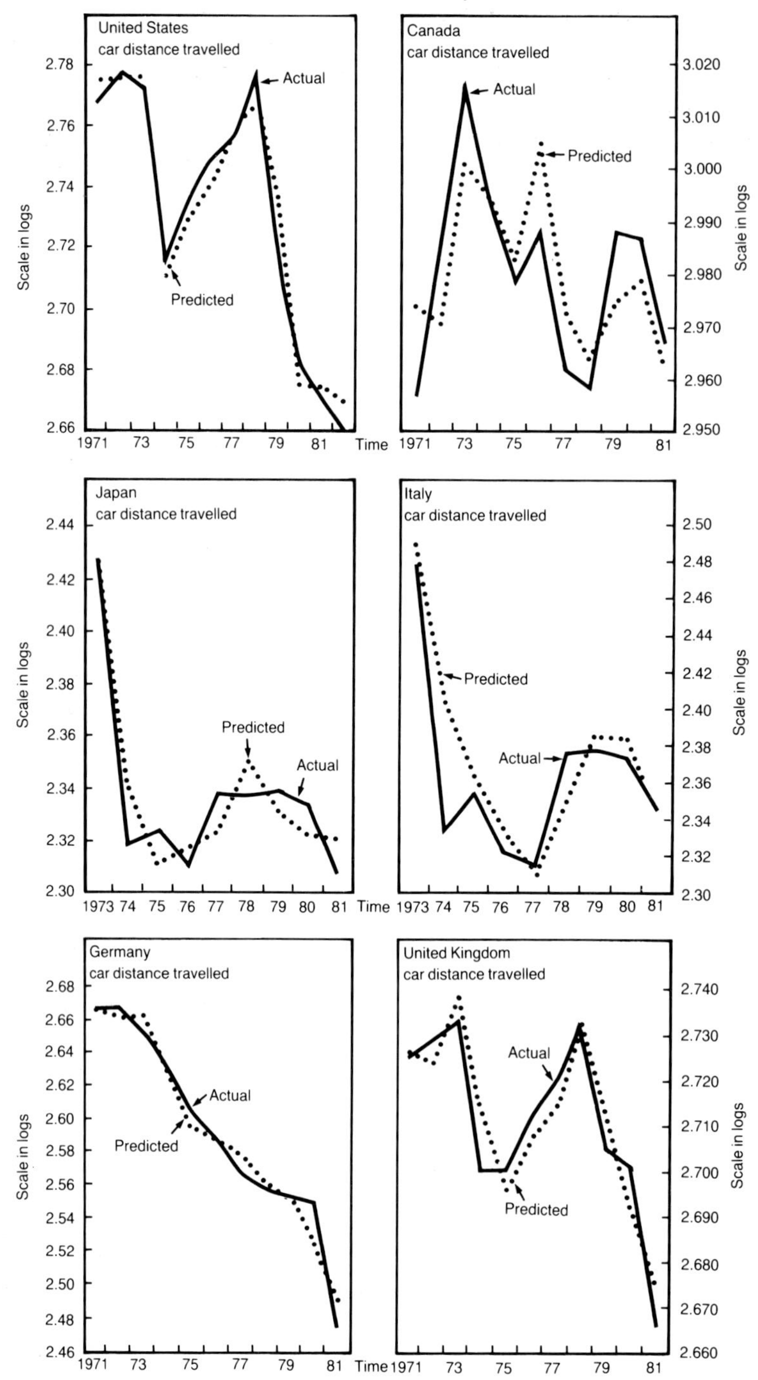

Fig 2.8 *Predictive accuracy of elasticity models*
(Source: IEA, 1984)

TABLE 2.7
Road traffic volumes: OECD and Great Britain (Billion veh km)

(A) Passenger Cars	*1970*	*1975*	*1980*	*1985*	*1988*
Country/Area					
N America	1534.7	1789.0	1941.5	2180.0	-
Japan	120.6	176.0	241.5	275.6	-
Australasia	73.5	90.1	99.7	116.0	-
OECD Europe	855.5	1081.5	1288.7	1427.0	-
OECD Total	2584.3	3136.7	3571.3	4000.0	-
Of which Gt Britain*	177.9	208.0	246.0	283.0	343.5
(B) Goods Vehicles	*1970*	*1975*	*1980*	*1985*	*1988*
Country/Area					
N America	369.8	477.6	671.5	800.0	-
Japan	100.0	104.9	141.5	146.5	
Australasia	18.2	23.0	30.6	43.6	-
OECD Europe	177.6	200.5	239.9	267.7	-
OECD Total	665.6	805.9	1083.5	1258.0	-
Of Which Gt Britain +	22.6	23.7	25.4	26.6	32.2
(C) All Vehicles	*1970*	*1975*	*1980*	*1985*	*1988*
OECD Total	3249.9	3942.6	4654.8	5258.0	-
Of Which Gt Britain	200.5	231.7	271.4	309.6	375.7

* Car, motorcycles and LGVs
+ HGVs and buses/coaches

Sources: OECD (1988) Tables 3A & B; Dept of Transport (1990) Table 2.1 and 7.4

The results of this first analysis are shown as the upper curve in Figure 2.10 for cars (etc.) up to 1982, and then in the lower, dotted line to 1988. It can be seen that fuel consumption (litres/100veh.km) remained roughly constant from 1970 to 1979, and then reduced in an irregular way to 1988. The reduction from the peak in 1979 to the last available figures in 1988 is nearly 17%.

The reduction in fuel consumption is broadly consistent with the projections made by the Working Party on Fuel Targets (Department of Transport, 1984a), with improvements taking place around 1980 when more efficient new cars were entering the fleet in large numbers, stimulated by the oil price rises in 1973 and 1979. However, more recent monitoring of new car petrol consumption in the Department of Transport (CMEO, 1990) shows (in the lower curve in Figure 2.10) that the improvements are not being sustained. Fuel consumption, weighted to represent actual registrations, shows a steady fall from from 1978 to 1984, and then a levelling off. There is an indication that fuel consumption has increased in 1988, 1989, and 1990 from the lowest value in 1987. The reduction from 1978 to 1987 was about 19%. The reasons for the lower rate of improvement are complex, and may be due to a number of factors like the fall in petrol prices, and an upturn in the economy, leading to larger engined cars being bought. In fact,

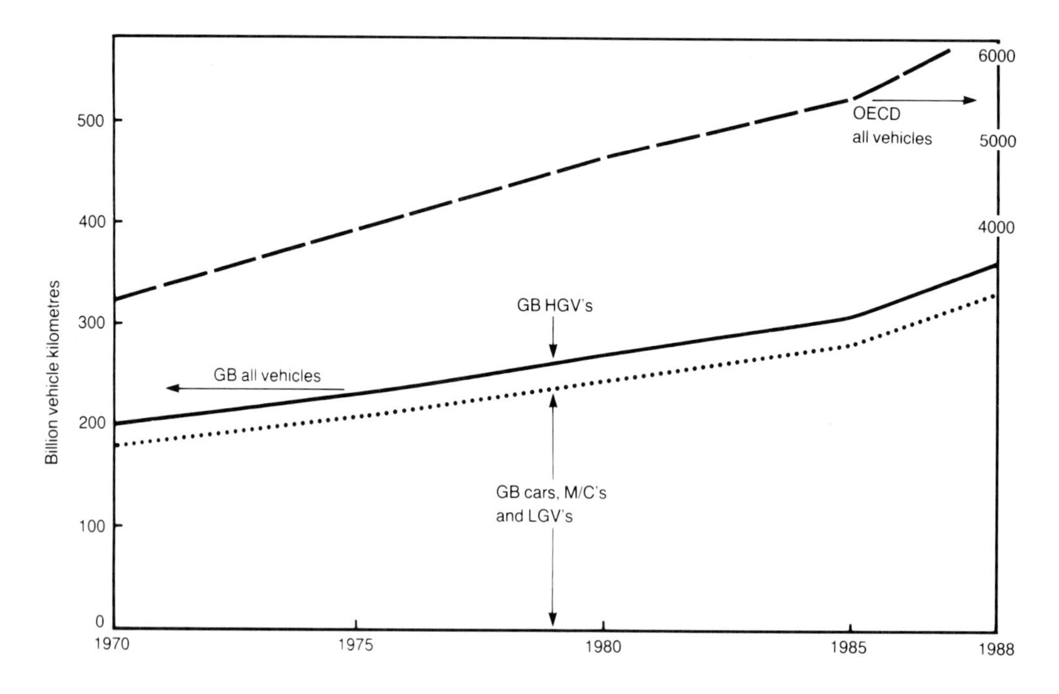

Fig 2.9 *Road traffic in OECD, and Great Britain, 1970 to 1988*
(Sources: OECD, 1988 and Linster, 1989)

between 1984 and 1989 the proportion of cars with engines of 800 - 1300 cc fell from 44% of new registrations to 36%, while the proportion of 1700 - 2000 cc rose from 13% to 32% (CMEO, 1990).

Before comparing the results for new petrol car models, and the trends derived from traffic and fuel use, it is instructive if an empirical attempt is made to remove the effect of including diesel powered cars and light vans from the traffic figures. In 1980, only 0.5% of the car and light van fleet was diesel powered. By 1983, the figure was 1.3%, and by 1988 had increased to 4.2%, with most of the increase concentrated in the private car sector. From 1983 onwards, the Department of Transport published statistics, year by year, which allow the numbers of diesel cars and light vans to be separately identified (Department of Transport, 1984b). By making reasonable assumptions about their fuel consumption and annual mileage, it is possible to derive "for example" values of traffic volumes and fuel used by the diesel vehicles. The approximate traffic volume for petrol cars and light vans can then be estimated. Table 2.9 sets out the relevant details. The resulting approximate specific fuel consumption values for petrol cars are shown as the upper of the diverging lines in Figure 2.10 from 1983 to 1988: the lower line shows the car fuel consumption without the empirical correction.

It can be seen that using the basic data (Table 2.8) would have exaggerated the improvement in car fleet fuel consumption from 1983 onwards by a significant amount. The amended figures show almost constant litres/100 km between 1983 and 1986. In 1987 and 1988, fuel consumption is reduced again. From 1979 to 1988, the overall reduction in fuel consumption is 11%, some 6

TABLE 2.8
Car and heavy goods vehicle fuel use and mileage

Year	Cars, motor cycles, vans			HGV'S, large buses, coaches				
	Motor Spirit M.tonne	*Billion veh-km.*	*litres/ 100 km.*	*Derv M.tonne*	*Billion Veh-km.*	*Billion tonne- km.*	*litres/ 100km.*	*litres/ 100 t.km.*
1970	14.235	177.86	10.84	5.035	22.59	82.2	26.4	7.26
71	14.964	188.72	10.74	5.186	23.24	-	26.4	-
72	15.899	198.78	10.84	5.254	23.71	-	26.3	-
73	16.927	209.27	10.96	5.658	24.74	87.6	27.1	7.65
74	16.484	205.83	10.85	5.518	23.95	-	27.3	-
1975	16.125	208.02	10.50	5.414	23.71	89.0	27.1	7.21
76	16.879	218.55	10.46	5.594	24.93	92.9	26.6	7.14
77	17.336	222.45	10.56	5.711	24.46	95.3	27.7	7.10
78	18.348	231.14	10.76	5.875	25.35	96.4	27.5	7.22
79	18.685	230.22	11.00	6.057	25.65	99.3	28.0	7.23
1980	19.145	246.00	10.55	5.854	25.43	89.7	27.3	7.73
81	18.718	251.91	10.07	5.549	24.42	90.2	26.9	7.29
82	19.247	259.93	10.03	5.731	23.95	91.1	28.4	7.45
83	19.566	262.91	10.08	6.183	25.10	92.3	29.2	7.94
84	20.226	276.94	9.90	6.755	26.14	96.6	30.6	8.29
1985	20.403	283.00	9.77	7.106	26.60	99.1	31.6	8.50
86	21.470	298.00	9.76	7.866	27.22	101.1	34.2	9.22
87	22.184	320.32	9.38	8.469	30.14	108.6	33.3	9.24
1988*	23.249	331.93	9.49	9.370	31.20	124.8	35.6	8.90
1988	23.249	343.50	9.17	9.370	32.23	124.8	34.5	8.90

Notes:

a) Figures for Motor Spirit and Derv delivered in the UK are a consistent series back to 1970.

b) Vehicle kilometres are the 1989 revision of traffic figures, and are consistent back to 1970. "1988*" indicates a provisional data set which has been revised in the more accurate 1988 figures. HGVs and large buses and coaches are added together to account for more of the Derv used, though diesel cars and vans are not included.

c) Tonne kilometres for HGVs are consistent 1978-88, which are unchanged from the 1976-78 series. Figures for 1975 and earlier years have to be adjusted for the change in definition of HGV in terms of gross weight instead of unladen weight.

d) Motor Spirit = 1355 litres per tonne
Derv = 1185 litres per tonne

Sources: Digest of UK Energy Statistics, Department of Energy, 1981, 1985, 1989a; Transport Statistics of Great Britain, Department of Transport, 1976, 1987, 1989 and 1990.

percentage points less than if no allowance had been made for the increased mileage than by diesel cars.

The tentative conclusion to be drawn about specific fuel consumption of the petrol car fleet is that consumption decreased sharply between 1979 and 1982, but then remained roughly constant until 1986, when a less marked reduction occurred. This reduction may not be sustained if the trends shown in the new model average are reflected in fleet performance.

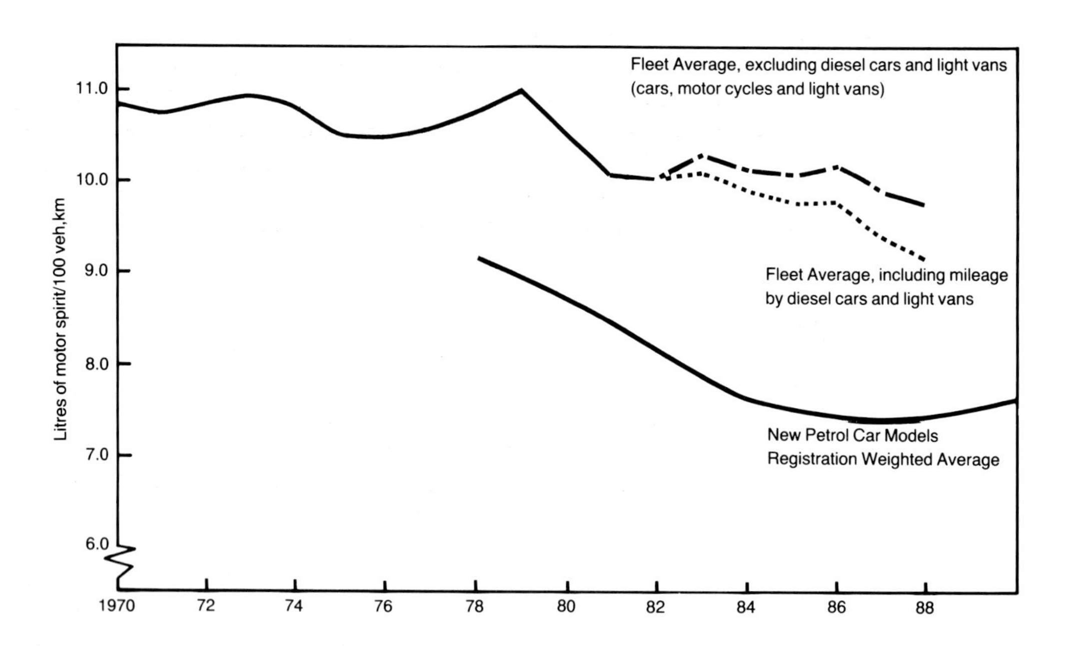

Fig 2.10 *Average fuel consumption of petrol vehicles*
(Source: Table 2.8, Table 2.9, Department of Transport, 1984b, CMEO, 1990)

TABLE 2.9
Estimated mileage and fuel use of diesel cars and light vans

Year	Diesel Cars *Number (000)*	*% of Fleet*	Diesel Light Vans (B.Veh.Km.) *Number (000)*	*% of Fleet*	Estimated Mileage *Cars*	*Light Vans*	Estimated Derv used (M.Tonne) *Cars*	*Vans*	*Total*
1983	69.	0.44	156.	9.66	2.04	2.24	0.14	0.17	0.31
1984	109.	0.68	183.	11.02	3.32	2.70	0.22	0.20	0.43
1985	166.	1.01	230.	13.49	5.06	3.40	0.34	0.26	0.60
1986	233.	1.37	286.	16.18	7.25	4.29	0.49	0.33	0.82
1987	311.	1.79	355.	19.47	10.19	5.64	0.69	0.43	1.12
1988*	402.	2.18	449.	23.04	12.88	7.15	0.87	0.54	1.41
1988	402.	2.18	449.	23.04	13.32	7.39	0.90	0.56	1.46

Notes and sources:

1. Diesel car and light van numbers from Department of Transport (1984b), and successive annual publications.
2. Diesel cars assumed to travel twice the annual mileage of petrol cars, and to have an average fuel consumption of 8.0 litres/100 km.
3. Diesel light vans assumed to travel the same annual mileage as petrol vans, and to have an average fuel consumption of 9.0 litres/100 km.
4. "1988*" indicates a provisional data set which has been revised in the more accurate 1988 figures.

This conclusion is only tentative because the results have been derived from data (on traffic and fuel used) collected for different purposes, and perhaps used beyond their limits of accuracy. The allowance for diesel cars is an obvious approximation. It is also known, for example, that revisions of traffic estimates are made between the publication of a provisional value one year, and its firm value next (see Table 2.8 for the change in 1988 figures). Major revisions of the traffic estimate series is not unknown. The last one was in 1989, and affected traffic estimates back to 1973 (Department of Transport, 1989). However, if provisional estimates are avoided, the trends in fuel consumption repay examination.

With the above provisos in mind, trends in fuel consumption of Heavy Goods Vehicles can be investigated in a similar way. Basic data is again given in Table 2.8, without allowance for the use of diesel fuel (derv) by diesel cars and light vans. (The mileage run by large buses and coaches is included with the HGV mileage to account for more of the derv used.) The results are shown, as per vehicle specific fuel consumption, in Figure 2.11. After 1982, the estimated amount of derv used by diesel cars and light vans has been subtracted from total fuel usage to give the lower line. The uncorrected values are shown as the upper line, and it can be seen that the effect of the approximate correction is quite marked. Taking the lower line as being nearer the truth, it can be seen that specific fuel consumption showed little change from 1970 to 1981, but that an increase then occurred up to 1986. It is possible that this increase was associated with changes in HGV regulations which allowed the use of heavier lorries from 1983 onwards. The changes may have been anticipated by operators, who bought the heavier vehicles before 1983, and ran them below design maximum weight. Between 1980 and 1988, all the increase in tonne kilometres of road freight was carried in HGVs over 25 tonne gross vehicle weight (Department of Transport, 1989). It is not, therefore, surprising that the per vehicle fuel consumption increased by 15% between 1981 and 1986. The most recent figures indicate that the increase may have been reversed.

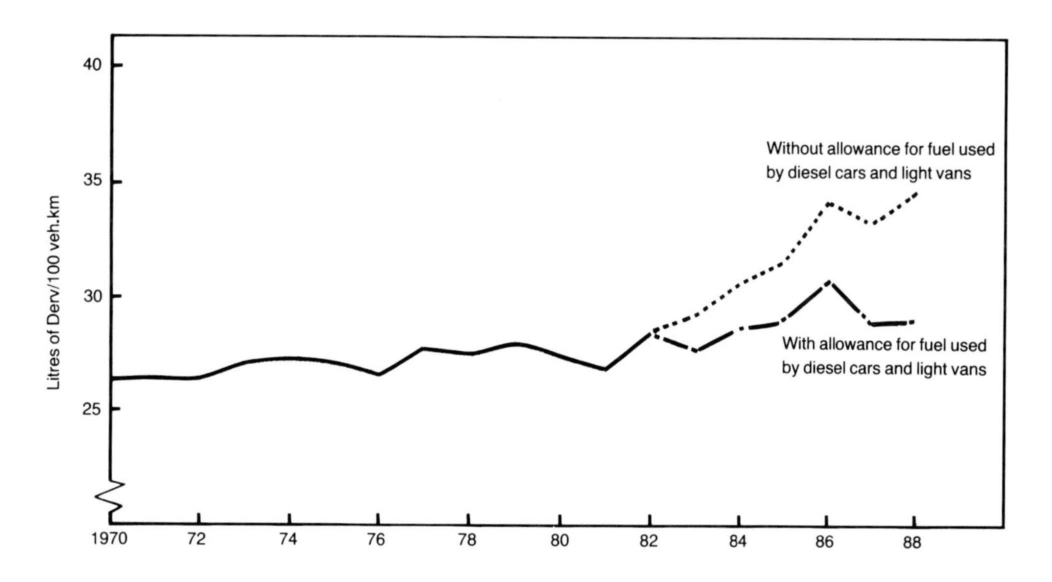

Fig 2.11 *Heavy goods vehicle fuel consumption (including large buses and coaches) (Source: Table 2.8 and Table 2.9)*

On the more rational measure of litres of fuel per freight tonne-kilometre moved, the approximate situation since 1970 is shown in Figure 2.12 for HGVs. (The approximation involves ignoring any payload carried by buses and coaches.) It can be seen that litres/tonne-kilometre hardly changed between 1970 and 1981, but then increased steadily to peak in 1986. The increase between 1981 and 1986 was 13%. By 1988, specific fuel consumption had reduced by about 9% using the lower curve, with allowance made for fuel used by diesel cars and light vans. The reasons for the changes are likely to be a complex mixture of operational factors, like lower load factors, larger vehicles, and possibly higher speeds on motorways increasing fuel use. The scope for vehicle design improvements was perhaps less than with cars, because HGVs are already powered by efficient diesel engines.

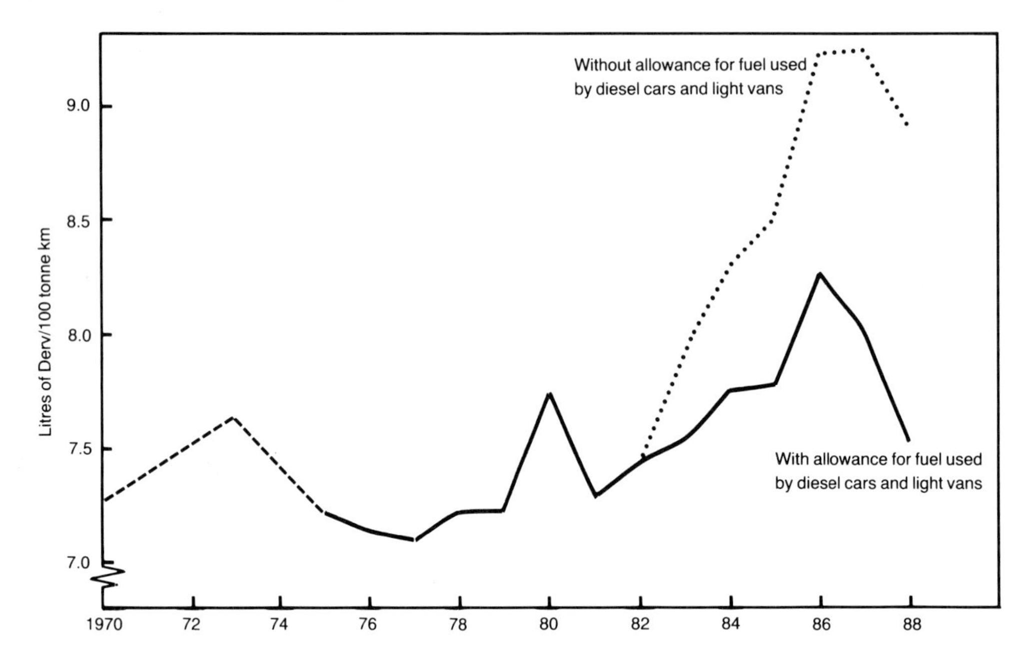

Fig 2.12 *Approximate HGV specific fuel consumption (Source: Table 2.8 and Table 2.9)*

2.5 Projected future use of petroleum products

Present day road transport needs an easily portable source of energy, which can readily be turned into the desired form - useful work. Liquid petroleum products from natural crude oil have fulfilled this need admirably for the first century of the internal combustion engine's dominance in road transport. But what are the forecast demands for fuel in the future, and for how long will natural crude oil be able to supply the demand?

The forecast demand will be examined first on the basis of statistics for Great Britain, as an example of a typical European country, and because the forecasts produced by the Department of Transport are easily accessible, and well documented. While any projections into the future are bound to be fraught with difficulty and uncertainty, there are advantages in knowing what assumptions have been made in the forecasting process.

For the purpose of this review, the key parameters are the traffic for all motor vehicles (in billion vehicle kilometres), and, as an example of vehicle stock, the number of private cars registered for use in the period in question.

For many years, the Road Research Laboratory (predecessor of TRRL) made major contributions to car ownership and traffic forecasts for the long term. The early forecasts of future numbers of motor vehicles in Great Britain up to the year 2010 (Tanner, 1962) were based on a logistic (S-shaped) curve, matching the growth in car ownership per head of population over the recent path, and becoming asymptotic to a "saturation" value in the very long term future. Later work expanded and elaborated on the original methodology, introducing additional predictive variables like economic growth and fuel prices (see Tanner, 1974; Tanner, 1977;); Projections of average kilometres per car as an input to national traffic forecasts were also made by Tanner (1981).

More recently the responsibility for making projections of motor vehicle numbers and traffic levels has rested with the Statistics Directorate in the Department of Transport. The latest forecasts (Dept of Transport, 1989) are new, and contain the following major assumptions:

(a) that GDP doubles (from its 1988 value) by the year 2025 for the low traffic growth projection, and more than triples for the high growth case.

(b) that retail petrol prices increase in real terms by about 60% for the low growth forecast, but by only about 20% for high growth by 2025.

The Dept of Transport forecasts to the year 2025, and the range between high and low growth futures, are shown in Figure 2.13 for numbers of cars registered, and in Figure 2.14 for traffic by all motor vehicles (except two wheelers). The actual values from 1970 to 1988 are also shown.

As an illustration of the strengths and weaknesses of long term projections, an early forecast for numbers of cars (RRL, 1965) is included in Figure 2.13. The forecasting method is basically the logistic curve (Tanner, 1962) modified by predictions of population growth in Great Britain. Bearing in mind that the car number forecast was published in 1965, it is interesting to see how the actual car stock fell below the forecast for most of the 70's and 80's, but that the forecast for the year 2010 is close to the latest "high growth" projection. It would be a mistake to read too much into the performance of one early prediction method, but it seems that the forecast of car population into the next century, as seen in 1965, was not greatly different from that based on 1988 experience.

In order to convert the estimates of future car ownership and traffic into fuel used for road transport, it is necessary to make some working assumptions about the changes in the fuel efficiency of vehicles far into the future. In very broad terms, the data in Table 2.8 shows that the overall fuel consumption of all vehicles improved from 12.6 litres/100 veh.km. in 1970 to 11.3 litres/100 veh.km in 1988. If a similar rate of improvement (around 0.6% per year) is extrapolated to the year 2025, fuel consumption overall would be 9.1 litres/100 veh.km - a 20% reduction from 1988. For the high traffic growth forecast, this would imply that the fuel used by road transport would nearly double, or would increase by 50% in the low growth case. If this did happen, it would mean that road transport alone would use between 60% and 80% of the petroleum products which are used today for all purposes. (In 1988, the proportion was about 40%.)

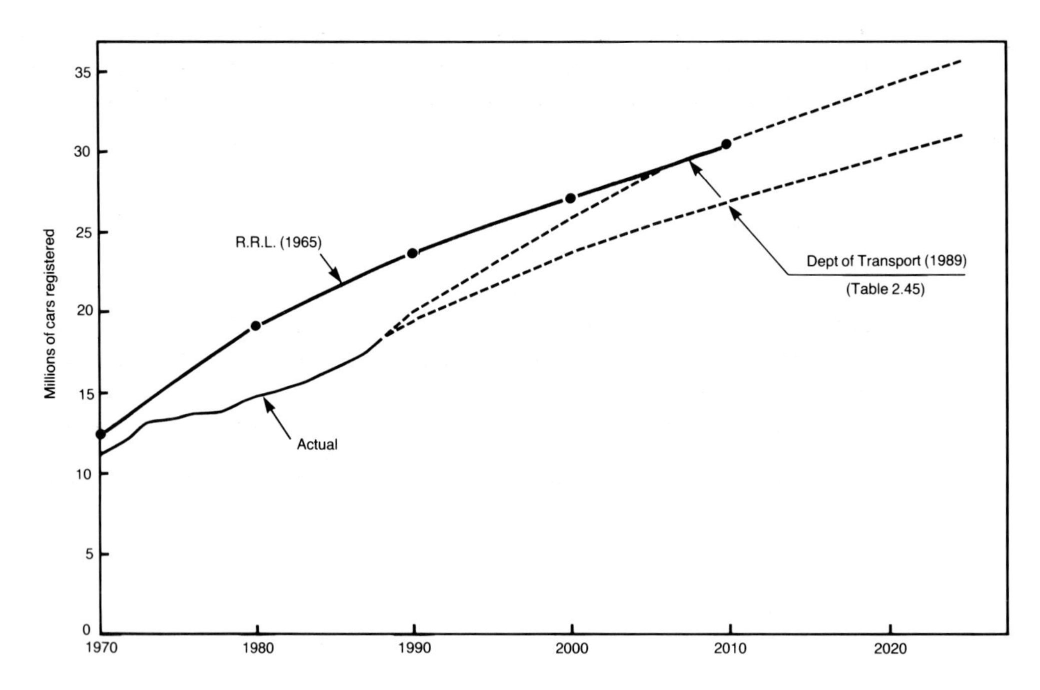

Fig 2.13 Actual and forecast numbers of cars in Great Britain

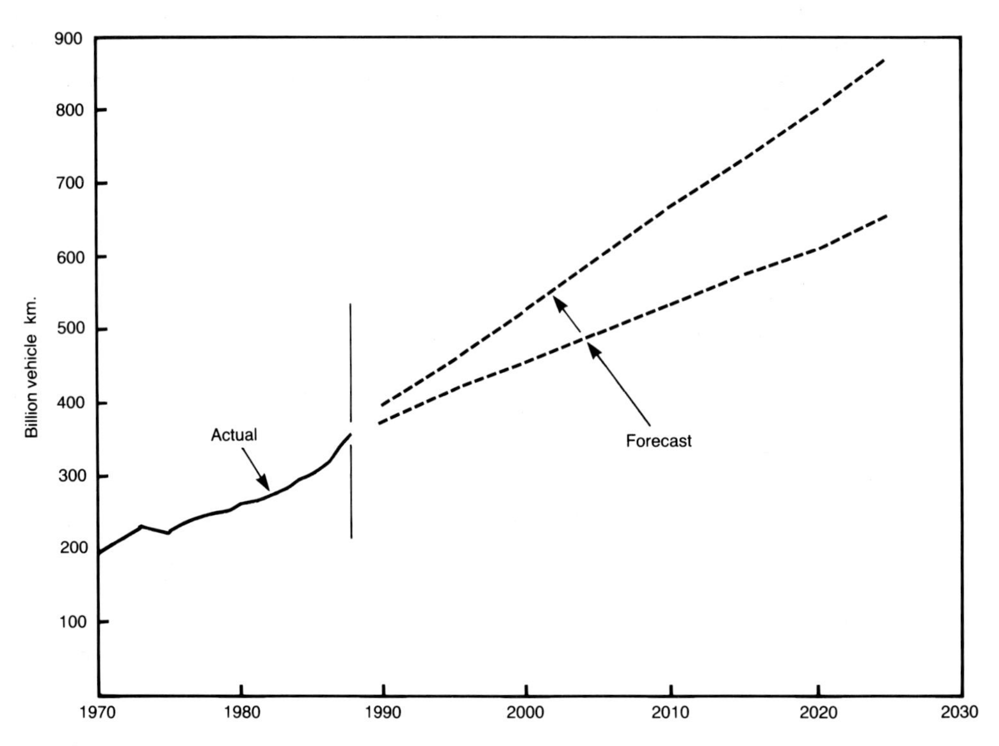

Fig 2.14 Actual and forecast traffic by motor vehicles (except two-wheelers) in Great Britain (Source: Department of Transport, 1989)

These estimates are, of course, illustrative and not intended to do more than set the scene for a brief look at the available reserves of natural crude oil. But they do indicate that oil products for road transport are expected to be a major part of the demand for oil well into the next century.

2.6 Oil reserves

Questions like "how much oil is there in the world?" and "when will the oil run out?" are easy to ask and nearly impossible to answer. Even if one rejects the theory (which has been seriously advanced by Gold, 1987) that crude oil and natural gas are seeping from massive reservoirs deep in the earth, rather than being the result of sedimentary action on marine life in the distant past, the total quantity which may exist as a result of prehistoric conditions is of academic interest only. What matters is the quantity available under certain economic and operating conditions. And as these conditions change, so does the quantity of "available" oil in different parts of the world. In the Middle East, the cost of crude fuel from on-shore wells is of the order of 0.2 US $ per barrel. From the North Sea, the cost was estimated in the mid-1970's to be from $3 to $9 per barrel (Cochrane and Francis, 1977), under the operating conditions which would have been thought impossible 40 years ago. So 40 years ago, the UK Continental Shelf would not have been included in world oil reserves because, by comparison with other fields, it was excessively expensive and difficult to tap. Changing prices and advances in technology of drilling and recovery have changed all that.

This example shows that at any particular time, published figures for oil reserves inevitably reflect the present, and projected, price (not cost) of oil, and the technology likely to be available to exploit the resource. Even the internationally accepted definition of oil reserves contains the time-dependent "economic" factor in all three categories of reserve:

(i) *Proven* - reserves which on the available evidence are virtually certain to be technically and economically producible (ie reserves which have a better than 90% probability of being produced).

(ii) *Probable* - reserves which are not yet proven but which are estimated to have better than a 50% chance of being technically and economically producible.

(iii) *Possible* - reserves which cannot yet be regarded as probable but which are estimated to have a significant (greater than 10%) chance of being technically and economically producible.

With these caveats in mind, remembering that reserves quoted for a particular year represent current thinking on what is "economical" and "technically producible", Table 2.10 shows the published remaining proved oil reserves, world-wide, at the end of December 1988. The well known concentration of reserves in the Middle East (62%) and in the OPEC countries (74%), together with their high Reserves/Production ratio* contrasts with the small reserves (and ratio) in, for example, both North America and the UK.

---

* High Reserves/production (R/P) ration indicates a long production life for the field.

TABLE 2.10
Remaining proved oil reserves at end of 1988

Area	*Billion tonnes*	*% of Total*	*R/P ratio (years)*
N America	5.5	4.8	10.1
Latin America	17.1	13.4	50.5
Western Europe	2.4	1.9	11.8
(of which UK:)	(0.570)	(0.5)	(5.0)
Middle East	77.3	62.3	>100
Africa	7.5	6.1	28.6
Asia & Australasia	2.7	2.3	17.5
Total N.C.W	112.5	90.8	50.2
Socialist Countries	11.3	9.2	14.5
TOTAL WORLD	123.8	100.0	41.0
(Of which OPEC)	(91.8)	(73.8)	(89.2)

Notes:
(1) Proved reserves are those in known reservoirs which can be recovered with reasonable certainty under-existing economic and operating conditions.
(2) Reserves/Production ratio (R/P ratio). The number of years the proved reserves would last at present annual production rate.
(3) N.C.W is the non-Communist world, that is, excluding the Socialist Countries. (See Table 2.1 for definitions.)
(4) OPEC is the Organisation of Petroleum Exporting Countries, presently:-
Ecuador, Venezuela, Iran, Iraq, Kuwait, Qatar, Saudi Arabia, United Arab Emirates, Algeria, Libya, Gambia, Nigeria, Indonesia.

Source: BP, (1989)

However, Table 2.10 is for proved reserves which, as already noted, might be expected to change with time. They have, in fact, remained almost constant over the past twenty years*, as new resources have been "proved" to balance the fields depleted by current consumption. The situation is illustrated in Figure 2.15 for proved reserves, and Figure 2.16 for R/P ratio. Another reason why remaining reserves, overall, tend to remain static in practice is that it is not profitable (in the commercial sense) for the oil companies to invest in exploration and drilling for reserves which will not be needed for some thirty years or more in the future.

For all these reasons, it is obvious that the total amount of oil remaining in place, world wide, is uncertain, but likely to be considerably more than the proved reserves as published at any time. For any particular oil field or region, the position is different because, as in any extractive process, the reserves in place at the start of production will eventually be exhausted. But even with a particular field, upgrading and redefinition of remaining reserves takes place over time, as can be seen by looking at the oil reserves on the UK Continental Shelf, as quoted between 1975 and 1989.

Table 2.11 (and Figure 2.17) show how the estimates of initial oil reserves in present North Sea discoveries have varied since 1975. While most of the increases at the "proven plus probable"

* Apart from a revision of Middle East reserves in 1987.

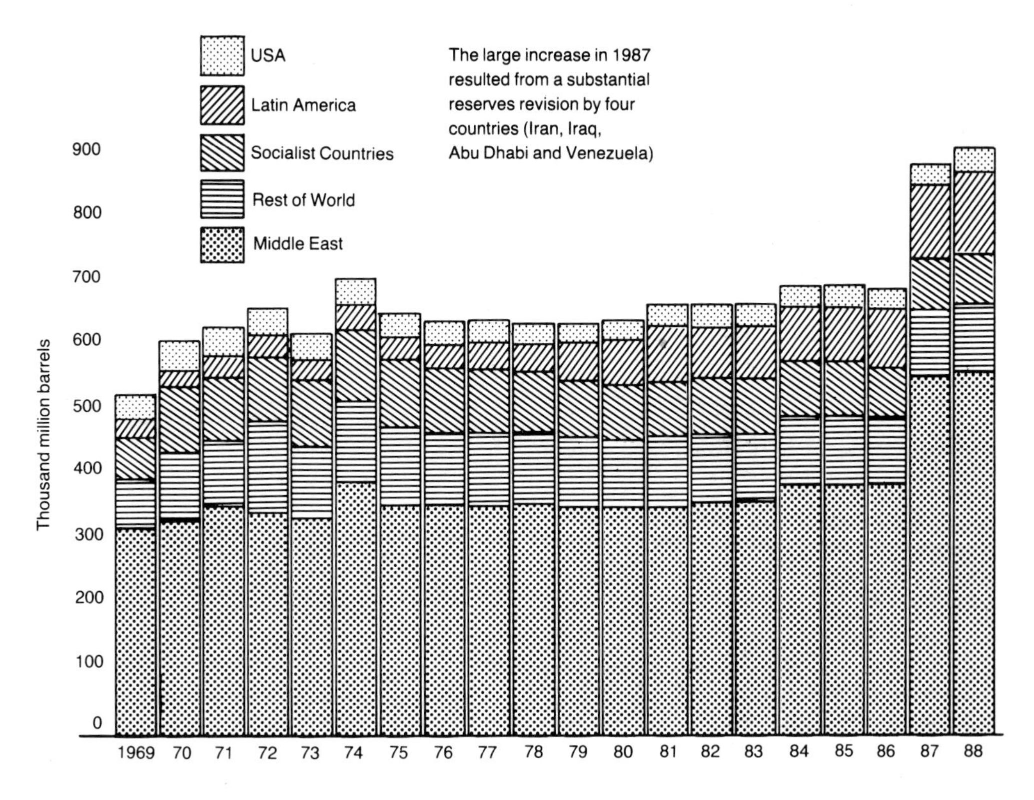

Fig 2.15 *Proved oil reserves 1969-1988*
(Source: BP, 1989)

level are the result of increases in producing fields, increasingly sophisticated reservoir management in some fields has led to significantly higher recovery factors than originally expected. Proven reserves have increased from 1350 to 1790 MTO between 1975, (when oil first came ashore) and 1989. In the same period, the maximum from "present discoveries" - the total of proven, probable, and possible - has increased from 2290 to 3090 MTO. When the cumulative production up to 1989 is subtracted from these initial figures, the range of remaining reserves is 510 to 1810 MTO, representing between 5 and 20 years production at present rates*. This is not a very long time to make, prove and bring into production new discoveries in the North Sea, especially as it is to be expected that new discoveries will be smaller and possibly more difficult to exploit than the original fields found in the early 1970's. The North Sea oil resource, as far as the UK is concerned, may be past its peak production. In a sense, 1989 was a watershed because just over half the proven and probable reserves in the present discoveries had been landed.

Nevertheless, the Department of Energy's "Brown Book", which is the annual report to Parliament by the Secretary of State for Energy (Department of Energy, 1990a), gave an optimistic picture of future development of North Sea oil (and gas) resources, commenting on

*Though the 1989 production of 91.8 MTO was considerably less than in the peak production year (1985) when it was 127.6 MTO.

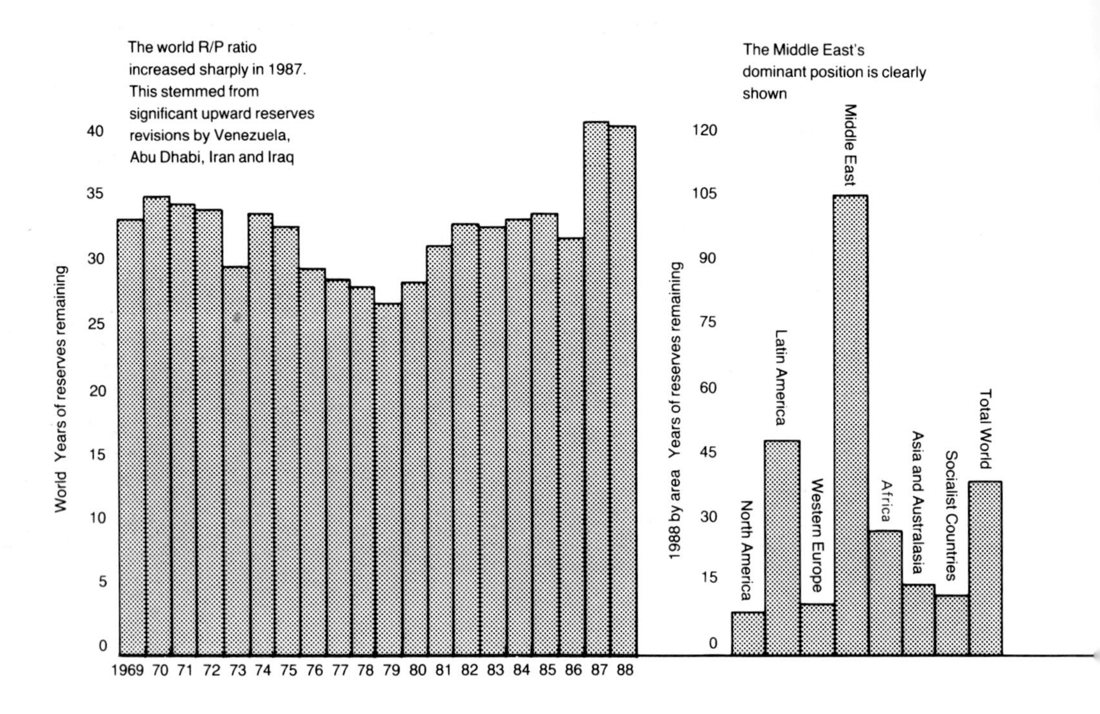

Fig 2.16 *World oil reserves/production ratio (Source: BP, 1989)*

TABLE 2.11

Estimated oil reserves on UK continental shelf

(Million tonnes of oil)	1975	*As at 31 December:* 1980	1989
Initial reserves in present discoveries:-			
Proven	1350	1390	1790
(Probable)	(580)	(575)	(690)
Proven & Probable	1930	1965	2480
(Possible)	(360)	(600)	(610)
Proven + Probable + Possible	2290	2565	3090
Annual Production of Petroleum	1.1	80.4	91.8
Total cumulative Production	1.1	263	1282
Range of remaining reserves in present discoveries	1350-2290	1125-2300	510-1810

Sources: Dept of Energy (1976); Dept of Energy (1981); Dept of Energy (1990b)

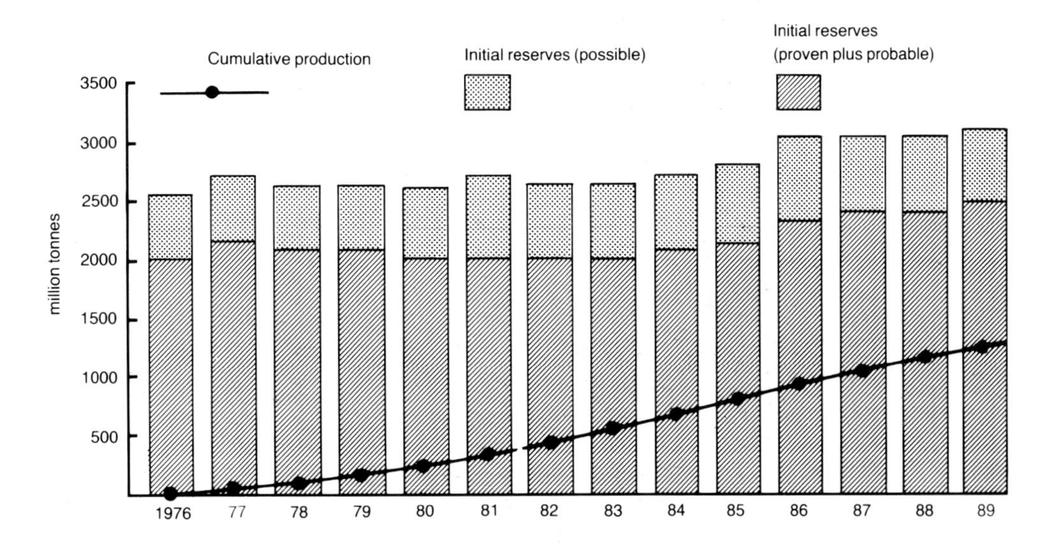

Fig 2.17 *Oil reserves in the UK continental shelf*
(Source: Department of Energy, 1989b and 1990a)

the high level of exploration and appraisal wells started in 1989, and the large number of awards made in the 11th Offshore Licensing Round.

The potential for future discoveries, as seen in early 1990, is set out in Table 2.12. The years of remaining resource, at 1989 production, is now increased to between 12 and 55 years, though of course the degree of uncertainty is greater than when the estimated reserves were based on discoveries already made.

TABLE 2.12
Estimated total potential of the UK continental shelf
(at 31 December 1989)

Million tonnes of oil initially recoverable:	
Discovered	1790-3090
Potential additional reserves	85-210
Undiscovered	530-3140
Total *initially*	2405-6440
Total *remaining* reserves	1123-5158

Source: Dept of Energy (1990a)

This review of the reserves of one particular geographical area shows how published figures have to be interpreted with care. For the UK's part of the North Sea, production can with certainty continue for 5 years at a rate of just over 100 MTO per year, and might be able to continue for 50 years, if oil thought to be there, but at present undiscovered, can be brought ashore. A cautious judgement might be that North Sea oil will continue to be brought ashore at the rate of 100 MTO per year for between 10 and 30 years - a length of time which could be increased by a policy of restricting production for premium uses (like petro-chemicals and road transport).

The conclusion about the "life" of North Sea oil is that the UK is likely to remain self sufficient, and a net exporter, into the early part of the next century. But the projections of car ownership and traffic growth (Figure 2.13 and Figure 2.14) already extend further into the future, and suggest that most of current oil production could be required then by road transport alone. Clearly some consideration needs to be given to the long term possibilities of alternative fuels for free ranging transport, and to their environmental implications.

2.7 Alternative fuels - an introduction.

A considerable literature exists on the pros and cons of different alternative sources of energy for general use, and for road transport in particular. It is not the intention here to go into all the possibilities, but to comment briefly on some of the "front runners" as reviewed by Langley (1987). His schematic summary of the main options for synthetic fuels for road transport is reproduced as Figure 2.18. This shows the basic feedstock, the conversion process, and the synthetic transport fuel which results.

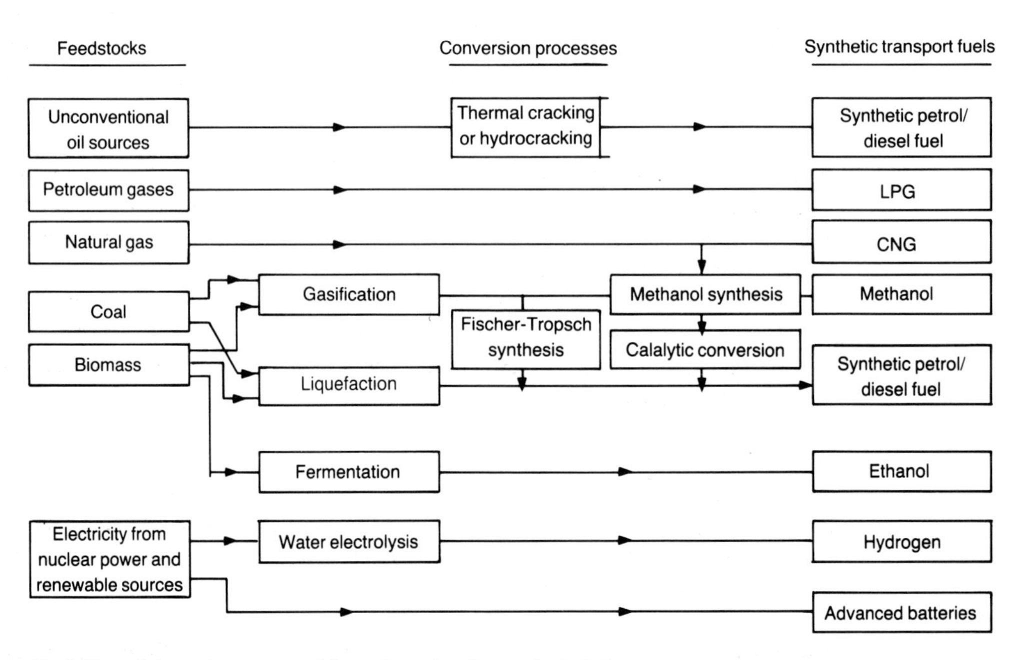

Fig 2.18 Schematic summary of the main options for synthetic fuels (Source: Langley, 1987)

To summarise the main options:

(a) *Unconventional oil sources* include heavy crude oil, tar sands and oil shale. All these can be converted to products matching present petrol and diesel fuel specifications by means of existing thermal cracking or hydrocracking techniques.

(b) *Petroleum gases* (propane and butane) can be liquified at room temperature by maintaining them under pressure, and they can then be used as LPG (Liquified Petroleum Gas) in suitably equipped vehicles.

(c) *Natural gas* (methane) can be carried in high pressure cylinders and used directly as CNG (Compressive Natural Gas) or it can be converted into methanol. Methanol, in turn, can either be used direct as a transport fuel with some modifications to the fuel system of the vehicle, or it can be converted by a further process to synthetic petrol.

(d) *Coal* can be converted to liquid fuels indirectly by gasification followed by a number of synthetic routes to give either methanol or a synthetic hydrocarbon fuel (petrol or diesel). Alternatively, direct liquidification methods can be used to produce hydrocarbon fuels.

(e) *Biomass* is material of biological origin other than coal, oil or gas (for example, wood, grain or algae). It may be processed to methanol or synthetic hydrocarbons by routes analogous to the coal conversion routes. It may also be possible, by fermentation, to produce ethanol.

(f) *Electricity* can be produced from nuclear power or from a renewable source such as hydroelectric schemes, wind power or tidal sources. It is treated separately here from electricity produced by burning fossil fuels (coal, oil or natural gas) because it needs consideration as a transport fuel in its own right - possibly by storage in a suitable form like a lightweight battery. Electricity produced from fossil fuels is less likely to be an attractive transport fuel because the fossil fuels can be used directly to provide liquid or gas products for transport.

Examples of many of the options described above can be found in different countries. There are a number of development projects for extraction of oil from tar sands and shale in the USA and Canada. New Zealand is using its large reserves of natural gas to produce transport fuels as a substitute for imported oil. South Africa produces half its requirement for transport fuel from coal conversion plants. Brazil has a large national programme for producing ethanol for transport from fermentation of sugar cane. In most cases, however, the programmes depend on substantial government support in the interests of national policy. Without this support they would not be economically viable at current market prices of fuels.

The list of options is not, of course, exhaustive. For example, ammonia, synthesised from air and water, has been proposed as a fuel from non-fossil sources, but its cost in energy terms is high so that might only be feasible "if cheap electric power were available from nuclear reaction or other sources" (ACEC, 1977). Hydrogen is another long-term possibility, which may depend on the availability of cheap electricity.

Most attention for an alternative to oil in the medium to long term future has been concentrated on the production of liquids from coal. One reason is that world coal reserves are well distributed geographically (unlike oil reserves), and that the Reserves to Production ratio is measured in hundreds of years, rather than in tens of years as for oil (BP 1989). The prospects for various alternative transport fuels are discussed later in Chapter 6, where the special place of fuels with low or zero carbon content is emphasised by the present concern over global warming possibly caused by carbon dioxide emissions.

2.8 Summary

Total world fuel consumption has increased by more than 50% between 1970 and 1988 and, while oil is still the largest component, it is becoming concentrated in premium applications like road transport where there is no alternative in the short term.

In the UK, total fuel consumption was almost the same in 1970 and 1988, but oil accounted for much less of the total. The proportion of oil used for road transport has more than doubled.

Major price increases of crude oil in 1973 and 1979 have had only minor effects on the demand for petroleum for transport. In the UK this is partly because high taxation has reduced the influence of crude oil price changes and general price inflation has quite quickly reduced real price. Another reason, for passenger cars, is the high proportion of company cars compared with other countries. It is thought that company car users are less sensitive to price changes than drivers who have to pay their full fuel and other motoring costs.

There is evidence of reduced fuel consumption per vehicle for passenger cars (some 15% over the period), but an increase in the per vehicle figures for Heavy Goods Vehicles, which is partly due to permitted increases in maximum HGV weight.

The projected future use of fuel for road transport has been examined in the light of recent forecasts of car numbers and traffic up to the year 2025. It is possible that road transport alone could use up to 80% of today's annual oil consumption.

Reserves of natural oil are concentrated in the Middle East, and proven North Sea oil is already half used up. There are many alternative sources of fuel for transport when natural oil becomes scarce and expensive, but they have differing problems in forming a satisfactory substitute for natural oil. There is therefore a continued incentive for fuel economy in road vehicles, so that the best use is made of existing natural resources.

3 Road vehicle economy - technical factors

3.1 Introduction

The need for improved fuel economy for road vehicles has been seen to be driven by different forces. Making the best use of finite natural oil reserves is one: reduction of greenhouse gas emissions, which could increase global warming, is another. For the private motorist and goods vehicle operator, reducing the fuel cost element of road travel is a direct incentive to achieve reduced fuel consumption.

There are many ways of reducing the fuel used by road transport. This chapter concentrates on those design and technical factors which can be applied to the vehicle. The way the driver, traffic and road system influences how the vehicle is used is the subject of Chapter 4.

Finally, the design changes which are discussed here leave the vehicle very much as a practical passenger or load carrier, with a performance in traffic comparable with present vehicles. The kind of single minded emphasis on fuel economy "at any price" which produces "cars" capable of over 6000 miles per gallon in Mileage Marathon competitions (eg Shell UK, 1990) is not the subject of this Chapter. There are lessons to be drawn from the design of Mileage Marathon vehicles, but they do not have direct application to the car used for travelling to work, to go on holiday, or for the weekend shopping trip. Quite apart from the fact that they tend to be ultra-light single seaters, they achieve their very high fuel economy by unconventional driving techniques. The driver uses fuel in bursts of acceleration at full throttle (of a very small engine) up to about 30km/hr, and then coasts with the engine off down to about 10km/hr, when the acceleration stage is repeated. Clearly this kind of progress is not a practical proposition in real traffic, and, while Mileage Marathon "cars" are technically interesting, they are not in the main line of development for improving fuel economy in practical cars. However, they have value in providing a "marker" for the ultimate in fuel economy, and Appendix C gives a short account of the Shell Mileage Marathon, which takes place each year, together with the miles per gallon achieved by the winners.In the paragraphs which follow, the influence of engine design on fuel economy is discussed first, and is followed by consideration of transmission systems and vehicle design factors. The effects of more stringent emission controls are then examined, and the Chapter concludes with a review of future possibilities for more efficient use of fuel in road transport.

3.2 Engine design

Most road vehicles today are powered by the Otto cycle spark ignition petrol engine, or the Diesel cycle compression ignition derv fuelled engine. Both are reciprocating four stroke engines, working on a similar thermodynamic cycle. The ideal thermal efficiency of the cycle (useful work supplied per unit of energy in the fuel) is approximately:-

$$E = 1 - r^{-0.4}$$

where r = compression ratio

One major difference between the Otto and Diesel engine is that the latter works at a very much higher compression ratio, so that the theoretical efficiency is higher. A typical modern petrol engine may have a compression ratio of 9 or 10, compared with about 20 to 23.5 for a small naturally aspirated diesel. This would suggest that the ideal efficiency of the diesel would be about 20% higher, but much of this gain is lost by higher piston friction and other losses. On the other hand, the diesel engine cycle efficiency is also higher because more heat is transferred to the working fluid at constant volume (near top dead centre), rather than at constant pressure (during the expansion stroke) with the petrol engine.

Another important difference is the performance under part load. The advantages of the Otto cycle engine are its lightness, low cost and high power to weight ratio. But because it takes in a ready made mixture of fuel and air, to be ignited by the spark plug, control of power output is by "throttling" the intake mixture, so that pressure losses occur at part throttle opening. Low power output is thus obtained by reducing the effective compression ratio and lowering thermal efficiency.By contrast, the diesel engine has its fuel injected at high pressure direct into the cylinder. For reduced power, less fuel is injected per cycle, but the compression ratio remains unchanged, and there are no "throttling" losses. The diesel engine thus has theoretical (and indeed actual) advantages in efficiency and fuel consumption when running at less than maximum power output. As maximum power is used infrequently (for maximum acceleration, and perhaps top speed), the diesel engine has considerable fuel consumption gains over the petrol engine in real traffic conditions.

It also has disadvantages. It has a lower power output for a given size (cylinder capacity), is heavier (to withstand higher internal pressures), and is more costly to manufacture - partly because high pressure fuel injection equipment is more expensive than a carburettor for petrol/ air mixture.

While most road vehicles have Otto or Diesel engines, there are some with Wankel rotary engines, and some with reciprocating two-stroke engines. The Wankel engine seems unlikely to survive in a climate of increasingly stringent emission regulations, but there are developments of a petrol injection two-stroke engine, the Orbital engine designed by Ralph Sarich, which may have market potential in the near future (Autocar and Motor, 1989a).

Many reviews have been carried out over the years to see whether other engines are likely to break the present duopoly. Gas turbines (Brayton cycle), the Stirling engine, the Rankine (steam) cycle have all been assessed (see, for example: ACEC, 1979; Francis and Woollacott, 1981; Martin and Shock, 1989). The conclusion has always been that, even if the technical development were successful, the costs of these power plants would be higher than for present day engines. The investment needed for a change of power plant would tend to act as a commercial barrier to change, unless the advantages in reduced fuel consumption were very large and the fuel price very high. At present, even the advantage in fuel consumption of a diesel car has too long a "payback" period in fuel cost saved to attract the majority of private drivers (Waters, 1980), so the prospects for more expensive power plants do not appear good.

3.3 Comparison of petrol and diesel cars and vans

The fuel economy of the Otto cycle petrol engine has been elegantly treated in the classic book edited by Blackmore and Thomas (1977). No attempt will be made to go over the same ground here. Past TRRL work concentrated on comparing the fuel consumption of similar petrol and diesel cars in realistic traffic conditions.

The first trials, in 1978, used Volkswagen Golf cars, one with an 1100cc petrol engine, and the other with a 1500cc diesel engine (Weeks, 1981). This was the first design of lightweight high speed diesel engine for small cars which was competitive in terms of driver acceptance, engine response and noise with the equivalent small petrol engine (Hofbauer and Sator, 1977). The crucial cylinder head design was, incidentally, based on the Ricardo "Comet" indirect injection chamber designed in the 1930's (Ricardo and Hempson, 1968). The VW Golf led the way for other manufacturers to introduce diesel engined cars into their model range.

Before discussing the results of the trials, a digression is necessary to clarify a frequent source of confusion. Because diesel fuel has a higher density than petrol, some authors factor a volumetric measure of fuel consumption (eg, litres/100km.) to account for the fact that there is more energy in a litre of Derv - typically 9-10% more. This is only part of the truth. When energy used in the refinery process is taken into account, comparison of petrol and diesel fuel consumption on a volume basis is approximately equivalent to comparing primary energy consumption (see Weeks, 1981 and Francis and Woollacott, 1981). This is because more energy (fuel) is used at the refinery to distil or crack heavier oil to petrol than to produce diesel fuel. On this basis, the total fuel used (and CO_2 emissions), from the car and from the refinery should be compared using the volumetric fuel consumption. This was the situation in the early 1980's with the mix of petrol and heavier petroleum products appropriate to that time. If the demand for petrol fell dramatically because of a large shift in demand for diesel fuel, the relation would need to be looked at again. Competition from other users, like aviation, could also alter the balance in the future, because heavy fractions of crude oil might need to be "cracked" into diesel fuel. To balance this, however, the trend towards unleaded petrol has tended to make the refinery energy for petrol increase further. The situation is complex, but the current view is that the diesel car fleet could increase considerably before changes in refinery practice would be needed (in 1988, diesel cars were about 2% of the fleet). In the meantime, this review will continue to assume that diesel and petrol fuel consumptions, compared on a volumetric basis, in effect take into account the different refinery energy used to produce the two fuels.

Returning to the comparative trials, the two Golf cars had very similar road performance (acceleration and top speed), and shared the same body shell (see Table 3.1). They were driven on a radial route from Crowthorne to central London, and then round a circuit in the central area. Each car covered over 6000 km on test. The fuel consumption results are shown in Table 3.2. The diesel car saved at least 24% of fuel by volume under all conditions: in densely trafficked urban areas the saving was up to 40%. The fuel consumption comparison between the two cars is shown

TABLE 3.1
VW Golf petrol and diesel cars: basic characteristics

Type		*Golf LD (4 door)*	*Golf L (4 door)*
Fuel		*Diesel (Derv)*	*Petrol (2-star)*
Engine:	Capacity	1471 cc	1093 cc
	Compression ratio	23.5	8.0
	Max output (DIN)	37kW @ 5000 rpm	37kW @ 6000 rpm
	Max torque (DIN)	82 Nm @ 3000 rpm	80Nm @ 3000 rpm
	Max revs	5400 rpm	6300 rpm
Performance (Manufacturer's figures)			
	Max speed	141 km/h	139 km/h
	Acceleration 0-80km/h	11.5s	10.5s
	Acceleration 0-100km/h	18.2s	16.5s
	Fuel consumption (DIN)	6.5 litre/100km	8.3 litre/100km
Fuel consumption - ECE			
	Urban cycle (independent test for TRRL)	6.7 litre/100km	11.0 litre/100km

Source: Weeks, 1981

TABLE 3.2
VW Golf fuel consumption on different types of road
(With two passengers)

Route	*Diesel litres/100 km*	*Petrol litres/100 km*	*Ratio Diesel/Petrol (%)*
Central London	8.48	14.27	59.4
Crowthorne to London	6.19	8.91	69.5
Rural	6.01	8.69	69.2
Motorway	6.29	8.51	73.9
Average Mixed 50%/40%/10%	7.27	11.46	63.4

(Source: TRRL, 1980)

in Figure 3.1 for constant speed running. At different average speeds as constrained by traffic, the comparative consumptions are shown in Figure 3.2. The characteristic shape of these latter curves had previously been noted by Everall (1968). The high fuel consumption at low average speed is caused by the start-stop progression of traffic in congested areas, which wastes fuel by accelerating the car and then requires the brakes to dissipate the energy. Fuel used with the engine idling and the car stationary is also obviously wasteful.

In the Golf trials, the diesel car was probably an exceptionally economical design, whereas the petrol car was only average for its time. Some people felt that this was unduly favourable to the diesel car. There was also some adverse comment because only one vehicle of each type was used, though care was taken to check that, in fuel consumption terms, they were typical of other examples of the model (Weeks, 1979). With hindsight, the advantages of the diesel car were perhaps a little exaggerated, but the results provided a useful stimulus to encourage the wider use of diesel cars.

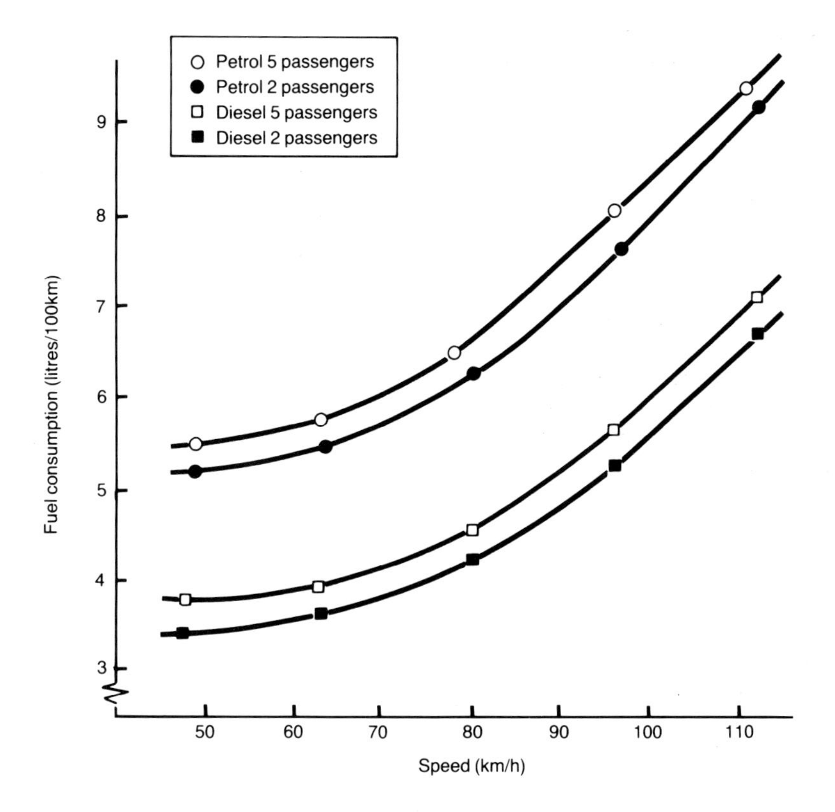

Fig 3.1 *VW Golf fuel consumption at steady speeds (Source: Weeks, 1981)*

Some years later, similar tests were carried out using petrol and diesel versions of the Vauxhall Cavalier. The work was done in 1985 by Redsell *et al* (1988) at the University of Loughborough. The main comparison was between a 1600cc diesel Cavalier and a 1300cc petrol Cavalier, though a 1600cc petrol version was also used as part of the concurrent investigation into the effects of driver characteristics on car fuel consumption. The 1600cc diesel and 1300cc petrol cars were not as well matched in road performance as the two Golfs. The performance of the petrol car was marginally better than the diesel, which had a lower top speed and poorer acceleration.

The results showed that in urban, suburban and motorway driving conditions, the diesel Cavalier used 22%, 17%, and 4% less fuel than its 1300cc petrol counterpart. Using a 40%/50%/10%* mix of these three driving conditions, would give an 18% fuel consumption advantage to the diesel car. The variation of fuel consumption with traffic-influenced average speed is shown in Figure 3.3a as regression curves for all three test cars. The results for the different volunteer drivers in urban conditions are shown in Figure 3.3b.

---

*See Chapter 4 for the basis for this proportional division, although Redsell *et al* use a 50%/40%/10% in Figures 3.3a and b.

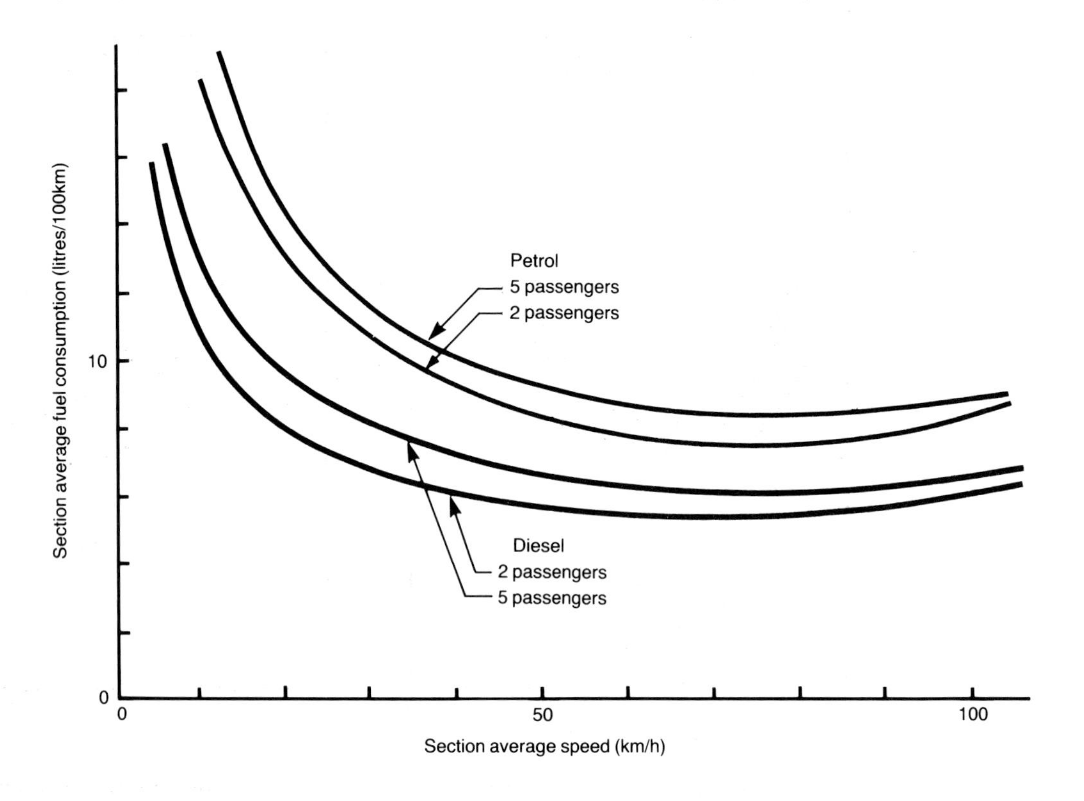

Fig 3.2 *VW Golf fuel consumption in traffic*
(Source: Weeks, 1981)

While 18% is a substantial saving, it is less than was found with the Golf cars. It probably reflects correctly the improvements in petrol engine economy that were achieved between the design dates of the two car types. On the other hand, there have been recent developments with small diesel engines which allow more efficient direct injection cylinder designs to be used (Motor, 1988), so the situation is not static.

Trials of a similar kind, but with less elaborate fuel measuring instruments, were undertaken using one tonne payload vans (Bedford Type CF) of the kind used for urban delivery duties (Wood *et al*, 1981). The results showed that, when corrected to a common average traffic speed (17.6 km/hr), the diesel van had an overall fuel consumption of 11.0 litres/100km, compared with that of the petrol van of 20.3 litres/100km. The diesel van thus used 46% less fuel at the low average speed of congested central London traffic. The variation of fuel consumption with traffic average speed is shown in Figure 3.4.

The conclusions from these tests are that modern high speed diesel engines in cars and light vans can lead to a substantial reduction in fuel consumption in real traffic conditions. The old perceived disadvantages of diesel-powered cars as noisy, smelly, and of low performance have largely been relegated to the past, and most major manufacturers now offer attractive and economical diesel cars, often using turbo charging to increase the specific power output of the

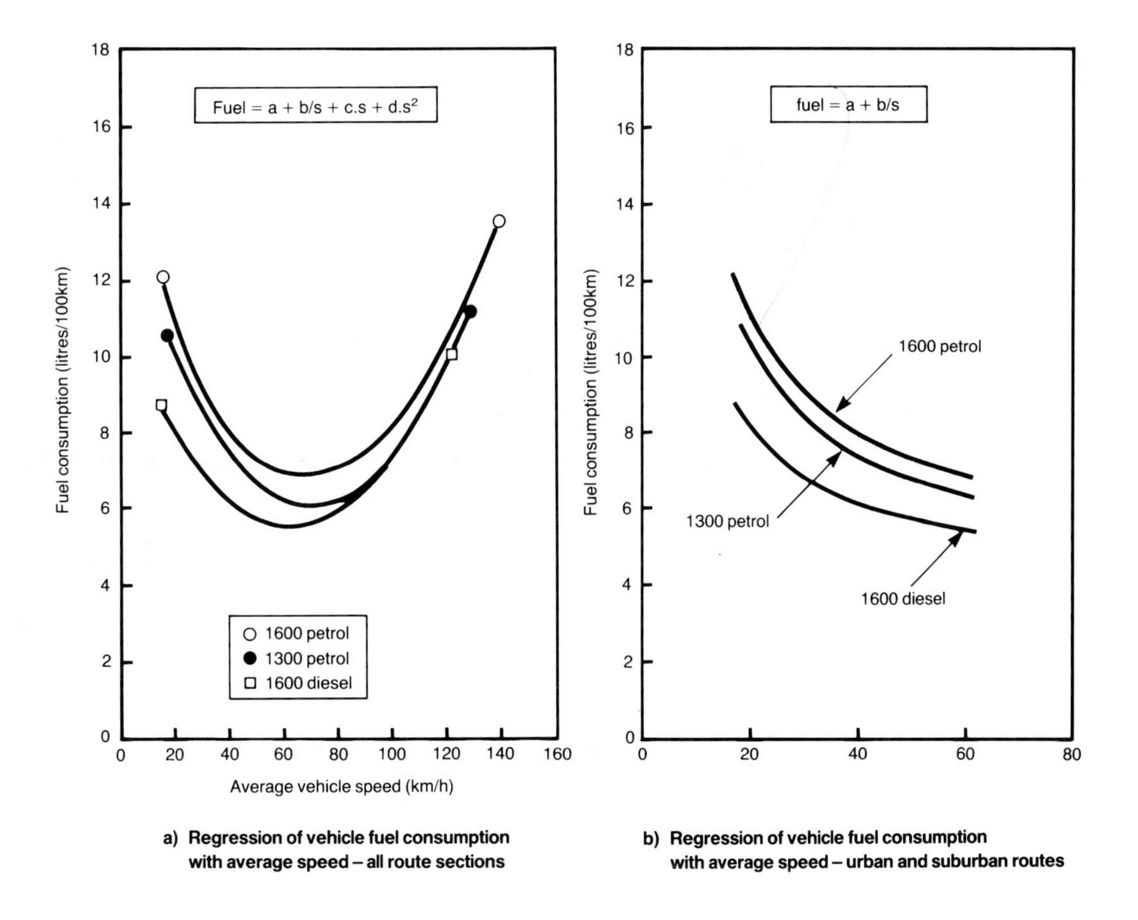

a) Regression of vehicle fuel consumption with average speed – all route sections

b) Regression of vehicle fuel consumption with average speed – urban and suburban routes

Fig 3.3 *Vauxhall Cavalier fuel consumption in traffic (Source: Redsell et al, 1988)*

engine. As an example, the Rover Montego saloon with a 2.0 litre turbo supercharged direct injection diesel engine has an overall fuel consumption (based on a 40%/50%/10% mix of the Official Figures) of 4.76 litres/100km (Autocar and Motor, 1989b). This is 30% better than the comparable performance Montego 1.6 litre petrol saloon. (Department of Transport 1990). While this comparison is based on regulatory test results, rather than real traffic experience, it does show the order of improvement in fuel economy that can be obtained.

3.4 Effects of running with cool engines

In all the trials referred to, measurement of fuel consumption was made with the vehicle's engine fully warmed up. But with private cars in particular, many short trips are made with a cold engine. It has been estimated that about one third of the national car fuel consumption could be saved if cold engines did not use extra fuel (Armstrong, 1983). This effect has been noted for petrol engined vehicles (Everall and Northrop, 1970: Blackmore and Thomas, 1977: Waters and Laker,

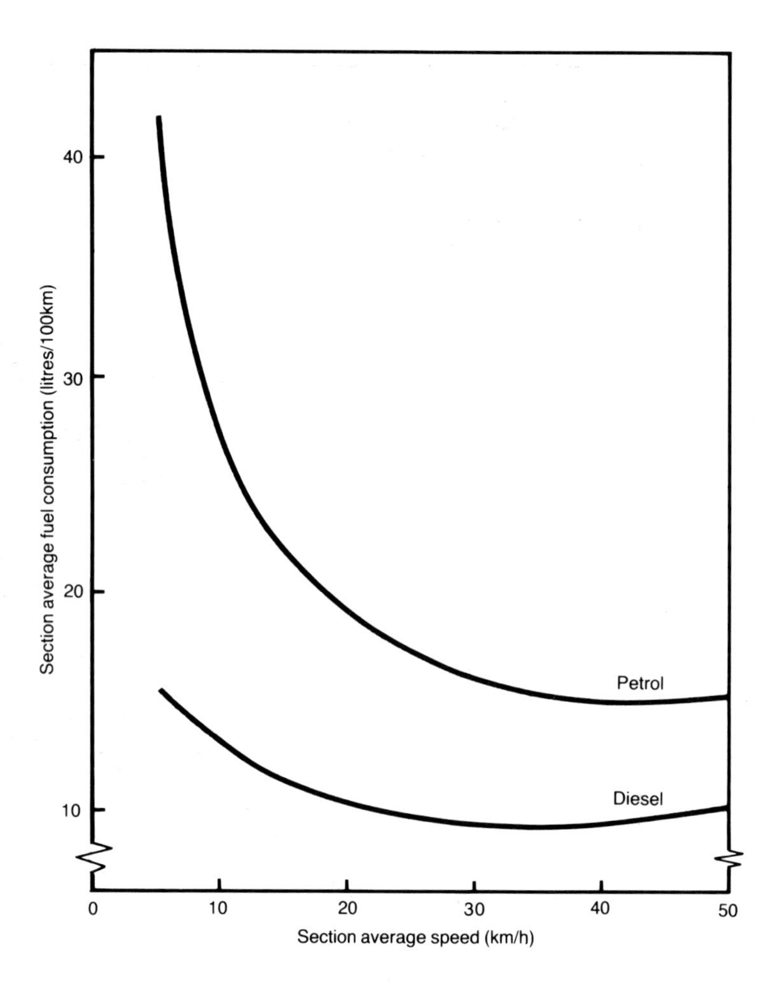

Fig 3.4 *Bedford CF van fuel consumption in Central London traffic (Source: Waters, 1982)*

1980) and also measured for a small diesel car in comparison with its petrol engined equivalent (Pearce and Waters, 1980). Tests made with the VW Golf cars showed that the diesel car had a substantial advantage in fuel consumption when trips started with a cool engine. One way of illustrating the extra fuel used was developed by Pearce and Waters in terms of the Fuel Equivalent Distance of a trip starting with a cold engine. The Fuel Equivalent Distance was defined as that trip length at the fully warmed up fuel consumption which would use the same amount of fuel as the actual cold start journey. The results, shown for the two VW Golfs in Figure 3.5, are taken from steady speed trials, but show that for each car they fall in a broad band not strongly dependent on speed. Thus, for the petrol car, a 5km trip with a cold engine is equivalent to a 13km trip with the engine fully warmed; for the diesel car it represents only about an 8km trip.

The conclusions drawn from the tests and illustrated in Figure 3.5 were that the diesel car suffered a "penalty" of only about 3km excess distance, compared with up to 11km for the petrol car. For

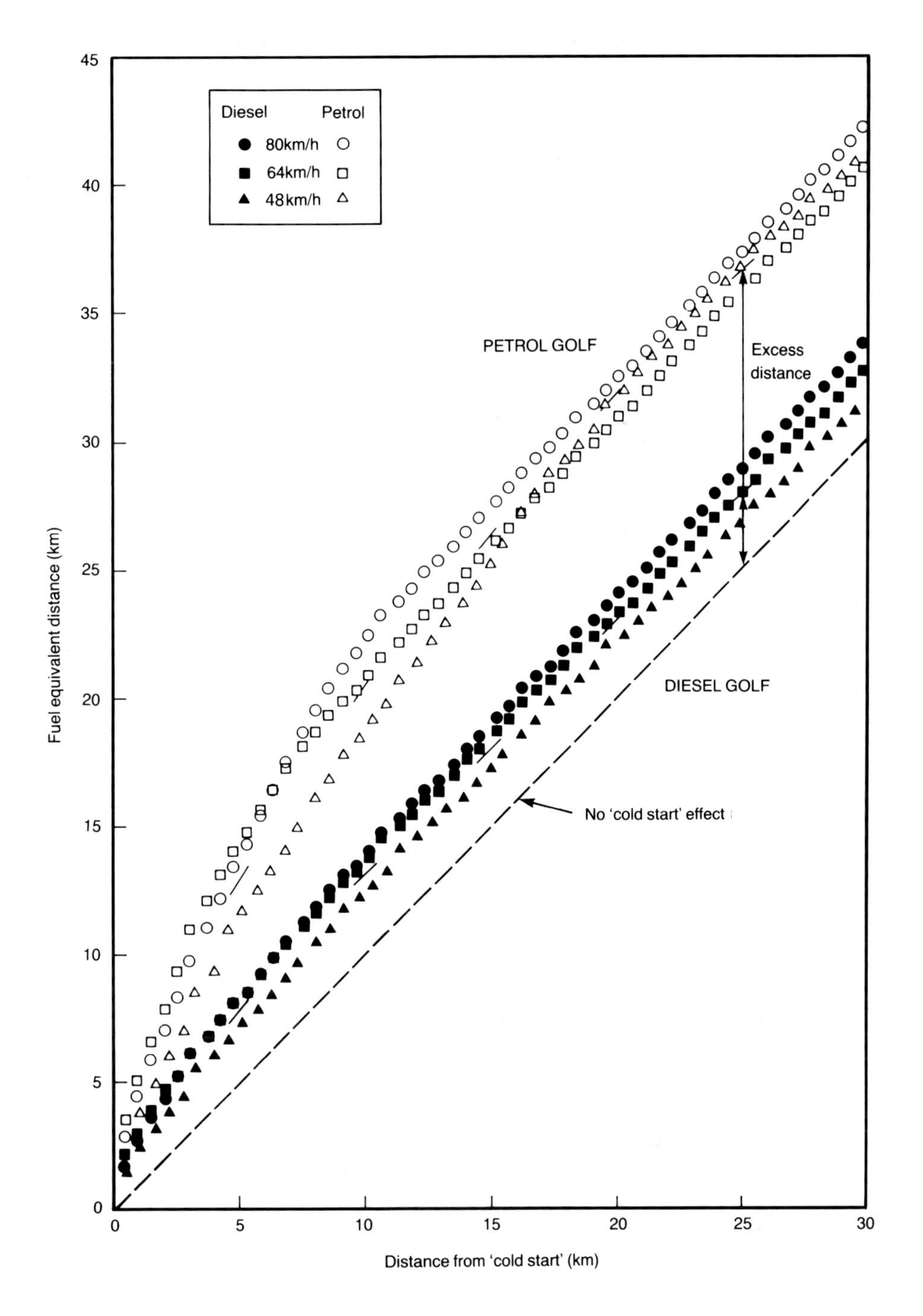

Fig 3.5 *The 'Fuel Equivalent Distance' for the petrol and diesel car (Source: Pearce and Waters, 1980)*

short journeys (5km or less) both cars use much more fuel from the cold start than when fully warmed up, but the diesel car still has an advantage. For the 5km journey used above as an example, the diesel would use 60% more fuel with a cold engine than when warmed up; the petrol car would use 160% more. Some of the advantage was because the petrol car was fitted with an automatic choke - a device notoriously profligate with fuel as Figure 3.6 shows. Other workers have found a smaller advantage for the diesel (OECD, 1982).

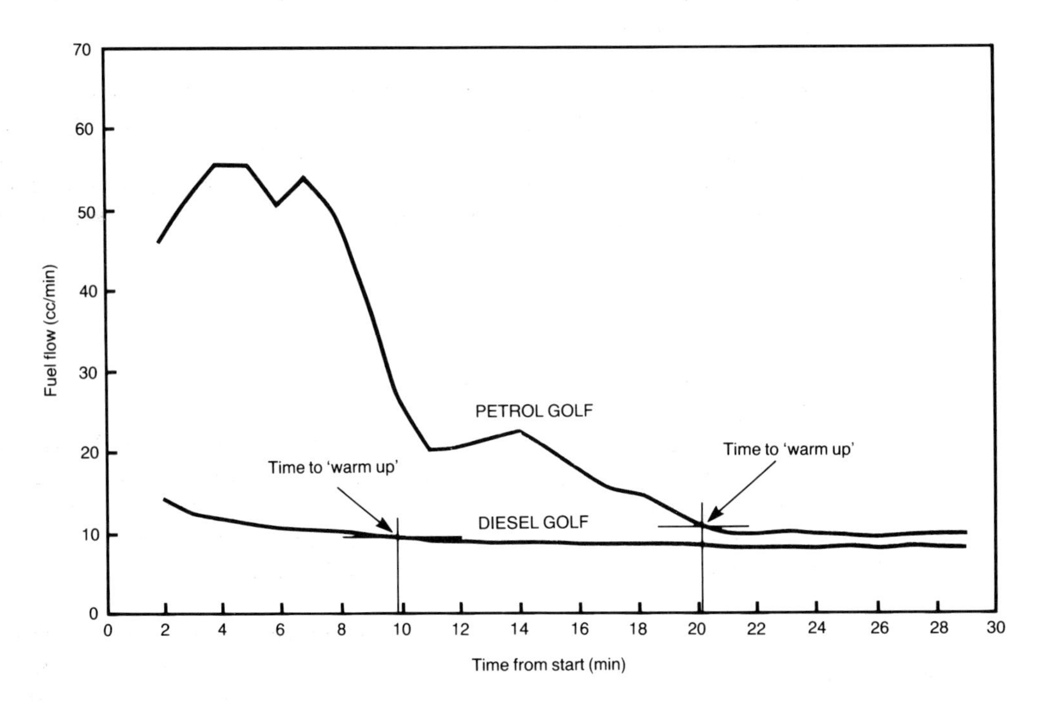

Fig 3.6 *Fuel flow while idling (3 point moving average)*
(Source: Pearce and Waters, 1980)

There are already indications that modern petrol engines, using fuel injection systems and electronic engine management, should be far better than the engine in the TRRL trials mentioned above which had a rather crude fuel enrichment device for cold starting. Design engine operating conditions, for both petrol and diesel engines, could be reached more quickly if the thermal capacity of the cooling system were lower. This is a feature of the Rover K series engines (Autocar and Motor, 1989c). VW have an experimental cooling system that uses the vapour from a small coolant reservoir as the cooling fluid (Autocar and Motor, 1990). Another approach is to use the heat generated by an engine during a journey which is normally dissipated during the time the car is parked. A VW experimental system stores this heat as latent heat in a tank containing a suitable compound so that it is available when the vehicle is used again. The system is claimed to retain most of the engine's heat over a weekend (Volkswagen, 1990). It would be useful to have detailed trial results with both the most up to date engines in current production, as well as for the newer developments described above.

One requirement to turn measurements on individual cars and vans into estimates of fuel saved country wide is the detailed pattern of trips of private cars, with time spent between journeys (when the engine can cool down) as well as the more usual trip length data. For cars, such information was available in the 1978/79 National Travel Survey (NTS) from travel diaries kept by survey subjects. The diaries were used by Armstrong (1983) in his estimate of the fuel used by cold running engines. A refinement, combining data from the travel diaries and the main N.T.S. to give the location and duration of car parking, was used by Watson *et al* (1986) in a study of recharging opportunities for electric cars. Updated information could usefully be applied to the cold engine question.

It is not only a cool engine that leads to increases in fuel consumption on a short journey. Williams *et al* (1985) showed in tests with a light van that losses in the transmission system, particularly the rear axle, were much higher when the vehicle started from cold. It was concluded that the resulting low transmission efficiency, combined with low engine efficiency and high (cold) tyre rolling resistance would go some way towards explaining why vehicle fuel consumption should be so high on short trips in cold weather.

3.5 Diesel engines for goods vehicles

Discussion of engine design has so far been restricted to petrol and diesel engines for cars and light vans, where petrol engines are by far the most common power source. Diesel engines are the most common type for HGVs, but the market share differs for the various vehicle classes. Diesel engines power over 50% of the heavy van population, over 90% of two axle rigid trucks, and 99% of all other rigid and articulated lorries (Martin and Shock, 1989).

In the design of a heavy goods vehicle, the performance characteristics of large diesel engines are matched with the different operating conditions, gross vehicle weight, and vehicle performance requirements of the operator. In one study on power train choice (Gyenes, 1980), it was found that optimum fuel utilisation was achieved by the choice of a comparatively small but highly rated engine. This was because of an improvement in engine operating efficiency, and an increase in payload capacity (because of lower engine weight). Charge cooled turbocharged engines of about 7 litres capacity at 32.5 tones GVW and 9 litres at 38 tonnes gave the lowest consumption at a fixed maximum power to weight ratio of 5 kW/tonne. But most of the fuel consumption reduction could be obtained by using a larger de-rated (turbo-charged) engine, and operators may prefer this choice because of the higher reliability of the less highly stressed power plant.Diesel engines for HGVs are already fuel efficient because the tax on fuel makes the cost of fuel around 25% of total operating costs (Newton, 1985), and operators of the heaviest vehicles are well aware of the fuel element in their costs. For the smaller goods vehicle, the fuel cost proportion is less, as Table 3.3 shows, and other matters may attract more of the operator's attention. While there is potential for further technical progress (for example, the more efficient "adiabatic engine"), there may, in the short term, appear to be more urgent areas for development like reduction of exhaust emissions (especially particulates), and further reduction of noise. Operators may feel that concentrating their management effort on improving load factors and reducing empty running may have quicker returns, and also reduce fuel use per tonne km. transported.

TABLE 3.3
Analysis of commercial vehicle operating costs

	Cost analysis (%) for		
	7.5 tonnes gross	*16.25 tonnes gross*	*38 tonnes gross*
Drivers' wages	53	48	33
Fuel	13	16	23
Maintenance/repairs	13	13	15
Depreciation	16	14	15
Vehicle Exercise Duty	1	5	8
Tyres	2	3	5
Insurance	2	1	1
	100	100	100

Source: Energy Efficiency Office, 1987, quoting the Freight Transport Association

3.6 Transmission systems

A variable gear ratio transmission to transmit engine power to the wheels is essential for petrol and diesel engined vehicles. Lightweight petrol and diesel engines will run over a speed range of perhaps 6 to 1 (1000 rpm at fast idle: 6000 rpm maximum), while large diesels for HGVs have a speed range of 3 to 1 or less. To provide power for acceleration and hill climbing, and to start with the vehicle stationary and go to maximum speed, requires a clutch and a transmission with a much larger speed range. Todays cars and light vans have, typically, a 4 or 5 speed gear box which effectively increases the speed range of the engine to around 30 to 1. Heavy lorries have an 8 or 9 gear transmission for the same reason.

The principles of matching engine and transmission for improved fuel economy can be most easily seen by looking at a typical power/fuel map - in Figure 3.7. This is for a petrol engine. Contours of constant specific fuel consumption (SFC) are plotted as a function of engine speed (rpm) and power output (BHP)*. The maximum power line (wide open throttle) shows the greatest power available at each engine speed. The solid line (marked "top gear") shows the power needed to run the car at steady speed on a level road. Each engine speed corresponds to a particular speed of the car on the road, depending on the gear ratio chosen. The power available above that needed for steady speed can be used for acceleration and climbing gradients.

It is obvious that the car in Figure 3.7 operates at low power up to quite high rpm (road speed), and that the specific fuel consumption - an inverse measure of engine efficiency - is high. The chain dotted line represents the power required if a higher gear ("overdrive") is used. Horizontal lines correspond to the same road speed in both gears, but the operating point in the higher gear has moved nearer the lowest part of the SFC contours, and engine efficiency is improved.

Much of the development of transmissions, including 5 speed (instead of 4 speed) gearboxes, automatic gearboxes with 4 instead of 3 gears, and continuously variable transmissions like the

* Fuel maps are also plotted with engine torque as ordinate instead of power, but this makes explanations more complicated.

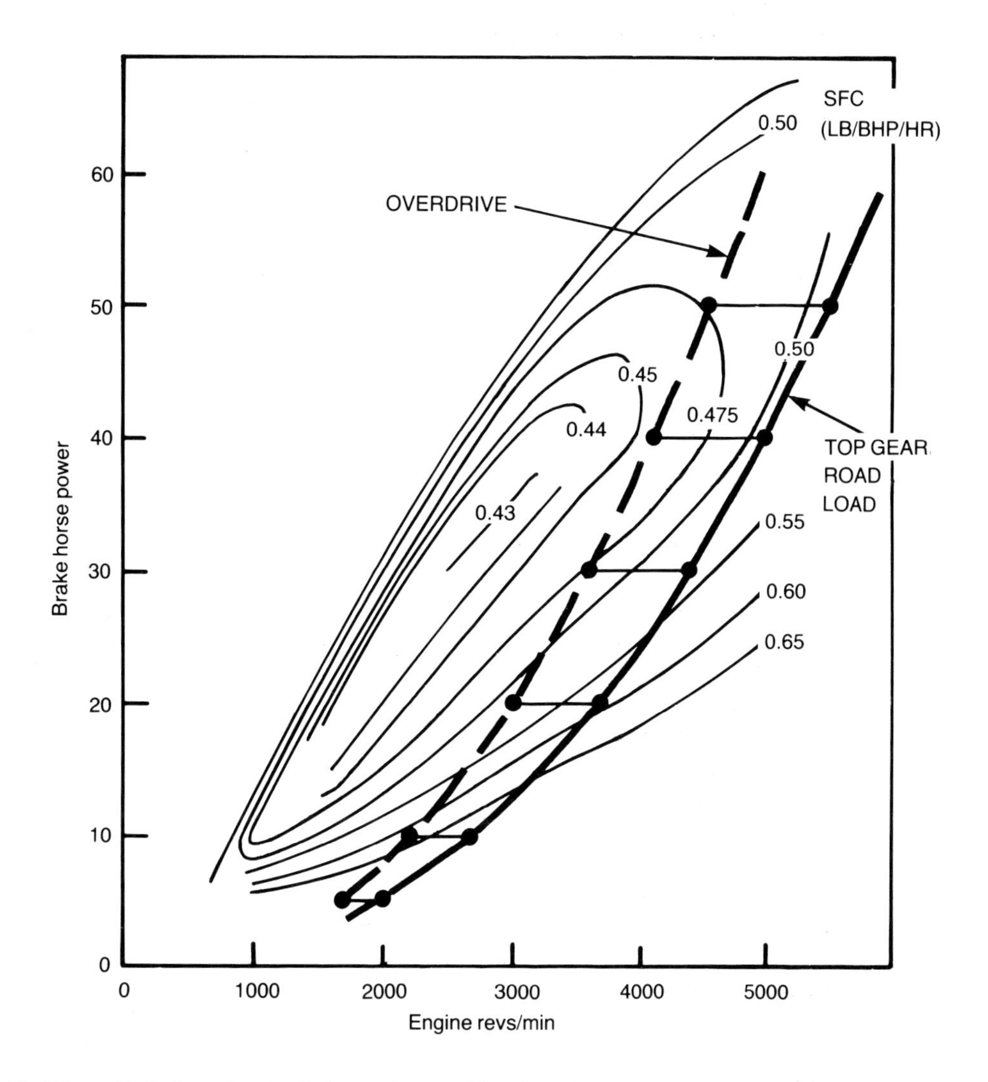

Fig 3.7 Typical petrol engine fuel map (showing effect of overdrive gear)

Perbury (Stubbs, 1980) has been aimed at improving fuel economy without making road performance worse. It is difficult to say how much improvement in car fuel consumption can be obtained by radical changes to the transmission, because it depends on the mix of driving conditions and the efficiency and speed range of the system. From computer modelling, Richardson (1980) deduced that, compared with a conventional 3 speed automatic gearbox, the ideal continuously variable transmission could produce a 30% improvement in fuel consumption with a given engine, while a 5 speed automatic could produce 20% in mixed urban and high speed driving. Other work has suggested that a 6 speed automated transmission could provide a 15% improvement in fuel consumption when compared with a 4 speed manual gearbox (Jarvis, 1984).

Computer modelling work (Ford, 1981) on a continuously variable transmission (CVT) compared with an equivalent car with a 3 speed automatic gear box showed gains which ranged from 3 to 37 per cent depending on the drive cycle used, and the transmission efficiency assumed. Highway driving showed slightly less of an improvement than urban driving, but, at the efficiencies then achieved in development, it appeared that gains in fuel economy of 14% to 18% were feasible. At the target efficiency of 91%, about 27-28% improvement with both driving cycles was predicted.

Stubbs (1980) also shows a reduction in fuel consumption over a wide range of speeds, from 45 km/hr to 120 km/hr, for a car with an experimental CVT. The improvement in economy is achieved because the transmission allows an optimum and variable high gearing to be used at all road speeds, instead of the gear ratio being a compromise between economy and performance, as with manual discrete gearboxes. In the experimental car, for example, a 15% reduction in fuel consumption at a steady speed of 110 km/hr was demonstrated, rising to 22% at 45 km/hr, with acceleration performance equal to a manual car with a 10% higher power/weight ratio. Stubbs points to other improvements that could be made, with further development.

One other development in transmission design which could lead to fuel saving is the "capture" of kinetic energy, usually lost when braking, by some form of regenerative braking system. The cost and extra complication of regenerative braking has to be balanced against the value of fuel and emissions saved, and such systems are most likely to be adopted in urban buses which make many stops, and operate in heavy traffic. Many demonstrations have taken place over the years, storing the braking energy in flywheels, air pressure vessels or electric batteries until needed to accelerate the vehicle again. Papers describing several systems have been presented by Bader (1981), Bauman (1981) and Hagin *et al* (1981). Some of the systems described and used experimentally were hybrid diesel/electric vehicles, but more recently interest has revived in diesel/pneumatic systems. While improvements in fuel economy as high as three times the original miles per gallon have been claimed (Slutsky and Levi, 1984), most developers see more modest improvements of around 30% (Nakazawa *et al*, 1987).

3.7 Vehicle design

The main vehicle design factors affecting fuel consumption are:-

- Mass (or weight)
- Rolling resistance (tyres)
- Aerodynamic drag (including the effects of cross-winds)

The increase in fuel consumption due to added passenger weight in small cars has already been seen in Figures 3.1 and 3.2. At constant speed, the main cause of the increase is the additional tyre rolling resistance and wheel bearing friction. In traffic (Figure 3.2), a low average speed implies start/stop driving, and the increased car weight requires more fuel to be used for acceleration, which is wasted later in braking. The effect of increased weight is therefore amplified at low average traffic speeds. This is why smaller and lighter cars are a powerful way of improving fuel economy. The effect is often called "down-sizing".

Down-sizing has a positive effect in improving fuel economy. In the USA, it has been said that much of the improvement required by the mandatory fuel economy regulations which raised the Corporate Average Fleet Economy* to 27.5 miles per US gallon (33.0 mpg) was achieved by down-sizing, and the penetration into the market of smaller foreign imports. However, there is concern over the difficulty of providing the same level of occupant protection in smaller cars (Reed, 1990).

The interaction of car size and occupant injury in crashes is extremely complex, and it is not easy to draw any but the most elementary conclusions from the data, which is far from adequate. However, it does appear that in single car accidents (eg a car hitting a tree or other solid object), car weight (or, more properly, mass) has no influence on occupant injury, though the physical dimensions of the car can be important. In frontal impacts, a larger car, with a longer crushable bonnet length, can reduce the deceleration suffered by the occupant and reduce injury. There is also obviously more crushable structure before the passenger cabin begins to be distorted. For this kind of accident, other things being equal, the larger car has an advantage.

In frontal collisions between two vehicles, the masses of the colliding vehicles affect their deceleration. When a small car crashes into a heavy lorry, the effect (on the car and occupants) is much the same as when hitting a solid object. But when a small car collides with a large car, both the relative masses, and size have an effect. Again, other things being equal, the occupants of the larger and heavier car will have less deceleration.

"Other things" are not equal, however, and early TRRL work on a proposed Crashworthiness Rating System (Penoyre, 1982) showed the difference in injury potential between cars of similar weight and size. Clearly the designer's skill and motivation for providing good occupant protection can play a part. This was emphasised by Folksam, a Swedish Insurance organisation (quoted by the Consumers' Association, 1988), who said, on the basis of accident injury experience that, in general, the risk of injury in different cars was closely related to the weight of the car - the heavier, the safer. But there were also big differences even between cars of similar weight, particularly with the smaller cars - the risk of injury in the Renault 4 was said to be roughly twice that in the VW Polo, for example.

This has been an over-simplified account of one possible disadvantage of relying too much on down-sizing to achieve improved fuel economy. The difficulty in designing satisfactory occupant protection increases as size (and mass) is reduced, especially if the mix of large and small vehicles in the traffic stream remains the same. Down-sizing cannot be rejected as a method of improving fuel economy, but this path may make the achievement of other important (safety) objectives more difficult, and the difficulties may increase as size and mass is reduced.

---

* In 1975, US Congress passed the Energy Policy and Conservation Act. Under the Act, a composite fuel economy value is calculated from test results intended to represent urban driving and highway driving. The composite is weighted at 55% urban and 45% highway. The Act requires that the average fuel economy of each manufacturer's total new passenger car fleet (not individual cars), commonly known as the Corporate Average Fuel Economy (CAFE), should meet minimum fuel economy values which increased in stringency with each new model year from 1978 to 1985. The 1985 value of 27.5 miles per US gallon represents an approximate doubling of new car fuel economy from the 1974 level of 14 miles per US gallon. (US Congress, 1980.)

With this reservation in mind, the sensitivity of passenger car fuel consumption to weight, rolling resistance, and aerodynamic drag was examined at TRRL by means of a computer model of a small car (Ford Escort). The model (Nowottny and Hardman, 1977) was used to produce the "trade-off" graphs shown in Figures 3.8, 3.9, and 3.10. These show the results for urban travel (represented by the ECE 15 urban cycle*), and travel at a steady 90 km/hr. More recent estimates have been made by other workers (and reported by Martin and Shock, 1989), and the same order of sensitivities were obtained.

With goods vehicles, the parameters are very different in magnitude, and TRRL has done a great deal of computer modelling and practical trials work to enable the effects on fuel consumption of the different factors to be understood. It is only possible here to show a very few illustrative results from a major research programme.

The computer simulation model (Gyenes, 1978a; Renouf, 1979) has been used for a number of investigations, and a comparison between simulation and experimental results for fuel consumption is shown in Figure 3.11. The simulation was further checked by estimating the fuel used by goods vehicles operating in the UK, and comparing this with estimates of actual use from diesel fuel tax revenue. Satisfactory agreement was obtained (Renouf, 1981). The model was used to examine the effect of power train choice (as discussed above), and also to examine the effect of gross vehicle weight (GVW) and payload weight on fuel utilisation (Gyenes, 1978b). Figure 3.12 shows one set of results for a 36 tonne GVW articulated lorry.

Experimental work on fuel consumption, and the techniques developed, have been described by Williams and Jacklin (1979), Simmons (1979), and Williams *et al* (1981). Results from trials on the rolling resistance of commercial vehicle tyres, and on the power requirements of articulated vehicles when cornering have also been reported (Gyenes *et al*, 1979 and Ramshaw and Williams, 1981). An example of the results from tests of aerodynamic devices to reduce drag is shown in Figure 3.13. Research has been carried out on models in wind tunnels (Naysmith, 1982), and the importance of testing at angles of yaw representing crosswind effects was emphasised (Ingram, 1978). At motorway speeds, and with a crosswind component giving a yaw angle of 10°, the drag of a typical HGV can increase by 70% (Williams *et al*, 1981). The results from wind tunnel tests and full scale trials have shown how the drag of the largest HGVs can be reduced. More recent wind tunnel tests on articulated lorry models have shown how their aerodynamic drag can be reduced, and pointed up similarities and differences in the required treatment for box-bodied and tanker vehicles (Garry, 1990a, 1990b).

Results of full-scale trials by one operator (TNT Ltd.) with side skirts, fill-in panels and a roof mounted deflector, all to reduce drag, have produced savings, as shown by fuel pump returns, of around 14% by the company fleet (Energy Efficiency Office, 1987). These tests have been followed up by a Demonstration Scheme, funded partly by the Department of Energy and involving Leyland Daf and Argos Distributers Ltd. to apply aerodynamic modifications to box vehicles in the range 6 to 24 tonnes GVW, and target savings of 13% per vehicle are aimed for (Energy Efficiency Office, 1988). During the trials, an average increase in fuel economy of 23% (mpg) was recorded for the 17 tonne GVW Leyland Daf Freighters (Energy Efficiency Office, 1990), and valuable operational lessons were learned.

---

*See Chapter 4.

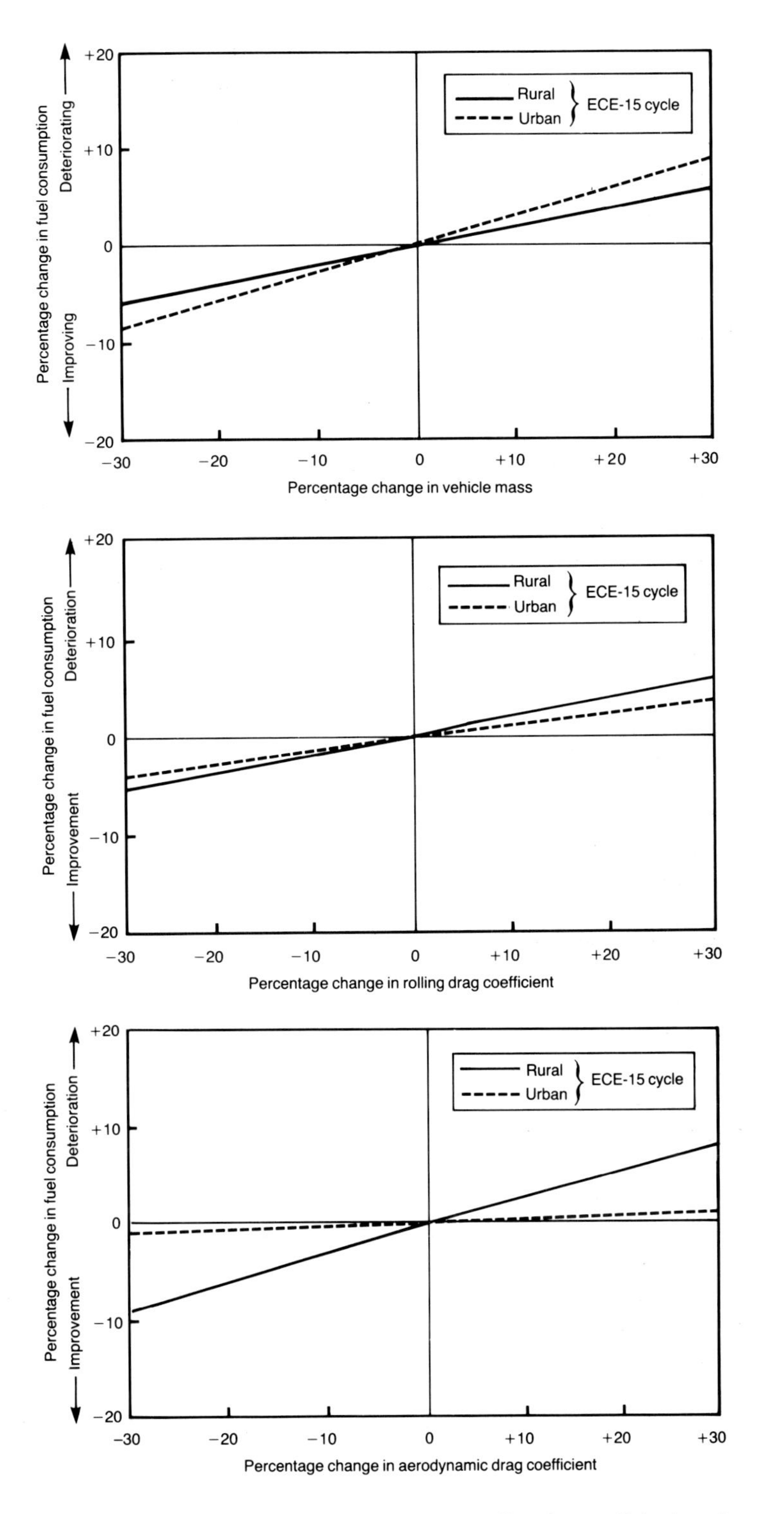

Fig 3.8/9/10 Computer simulation: effect of changing vehicle mass/rolling drag coefficient/aerodynamic drag coefficient (Source: Waters and Laker, 1980)

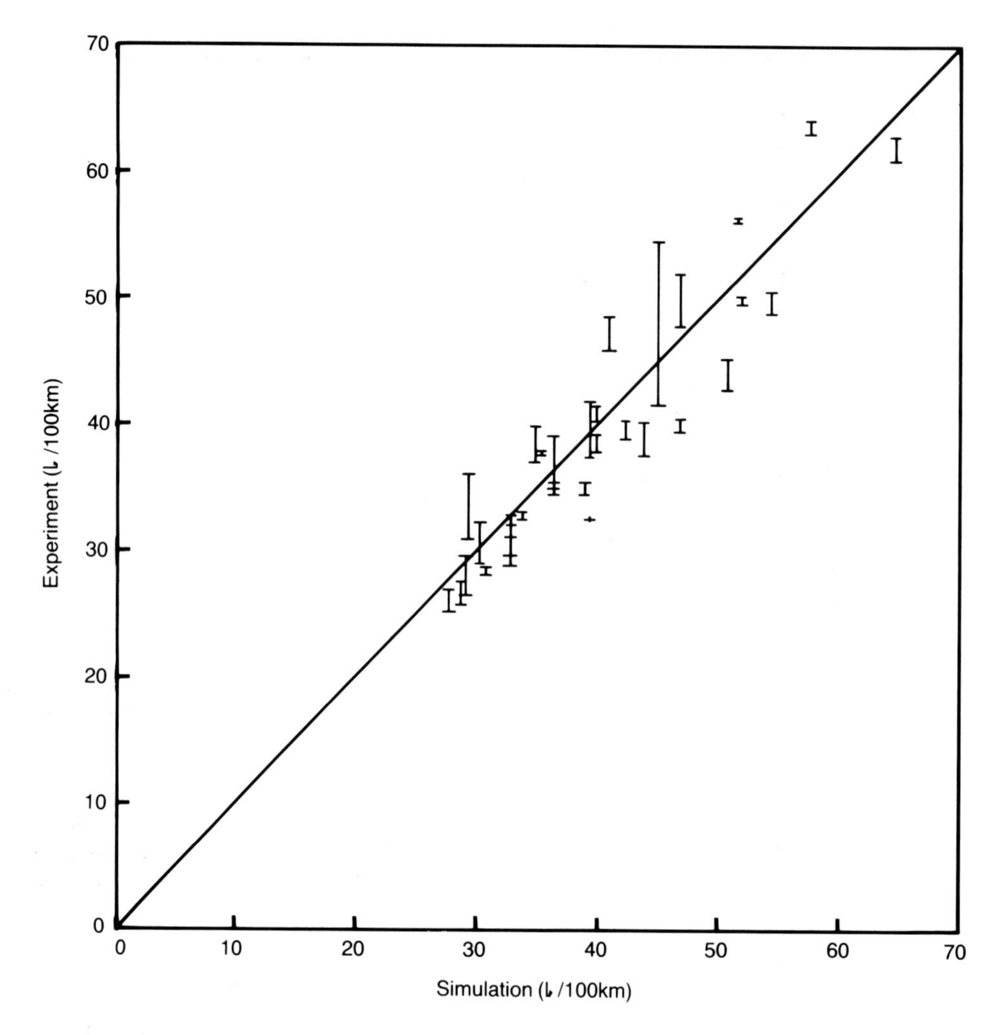

Fig 3.11 *Comparison between experiment and simulation for HGV's: 45° scatter diagram for all weights and road routes*
(Source: Renouf, 1979)

For the average articulated vehicle in the national fleet, Gyenes (1990) estimated the likely saving to be 10% to 13%, using a computer simulation. Trials with a modified articulated HGV have been carried out by TRRL for the Energy Efficiency Office, and a report is being prepared. The results confirmed the computer simulation estimate.

Many of the results from the TRRL commercial vehicle research programme were summarised in the booklet published by the Energy Efficiency Office (1987), and used to inform operators of good fuel consumption practice. The sketches in Figure 3.14 indicate how the aerodynamic shape of HGV's can be unproved.

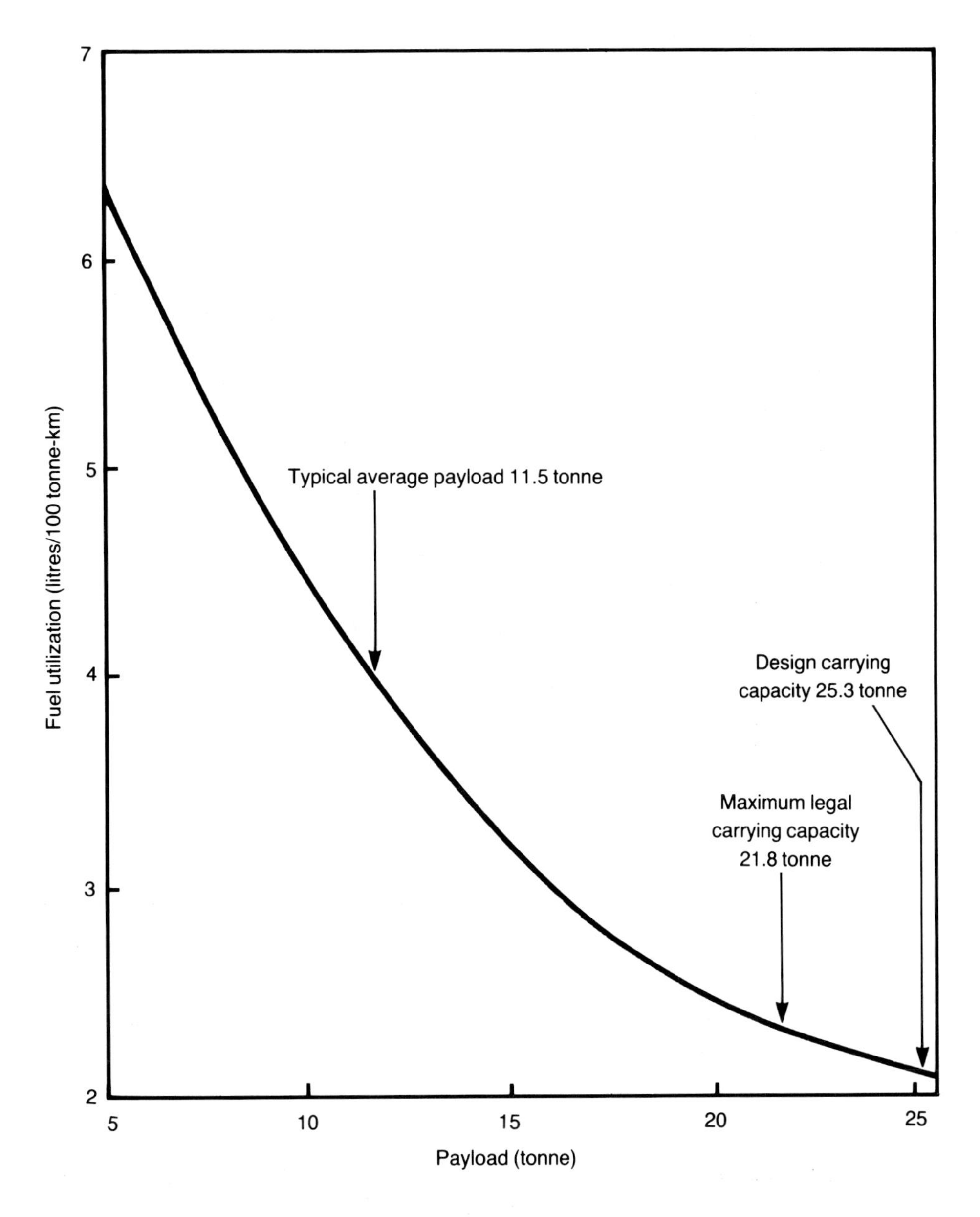

Fig 3.12 Fuel consumption variation with payload (36 tonne GVW articulated lorry) (Source: Gyenes, 1978b)

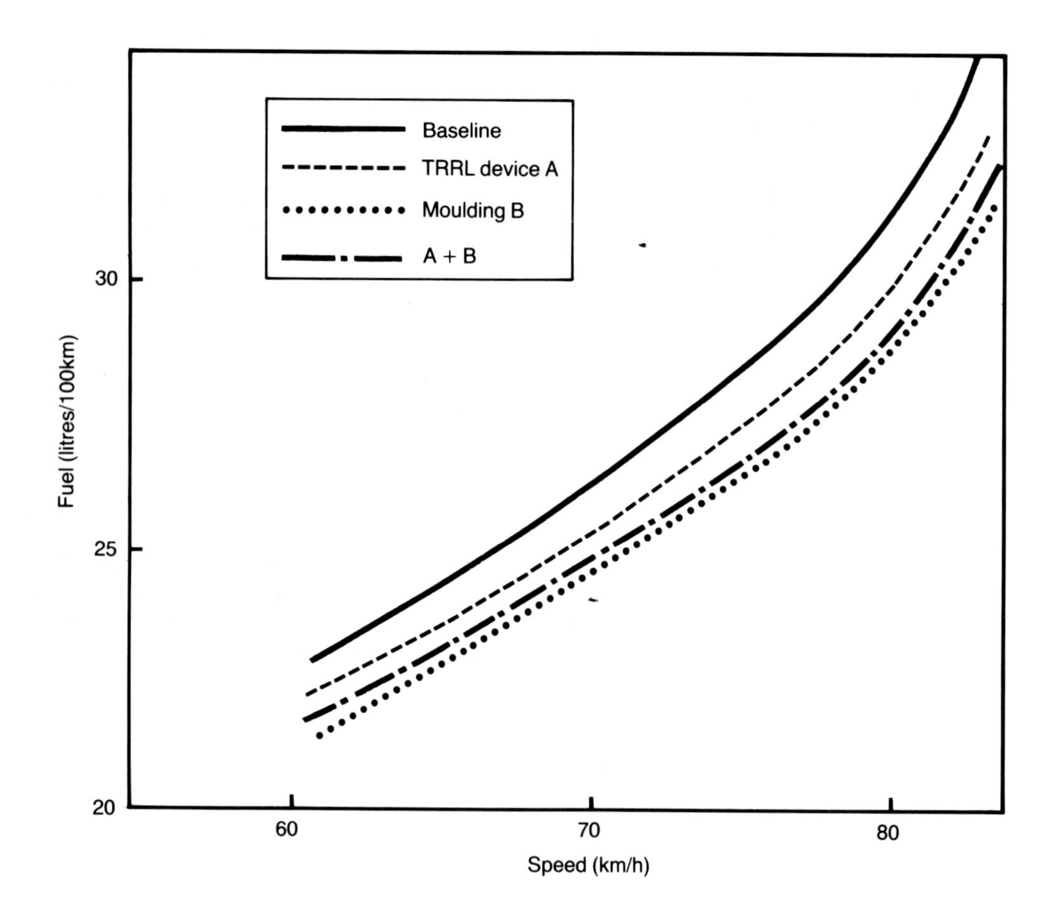

Fig 3.13 *Effect on fuel consumption of aerodynamic devices on a 16 tonne GVW 2-axle rigid lorry*
Notes:
(i) TRRL device A is a head board fitted above the cab and following the slope of the windscreen
(ii) Moulding B is a proprietary device fitted to the front of the box-body
(Source: Williams et al, 1981)

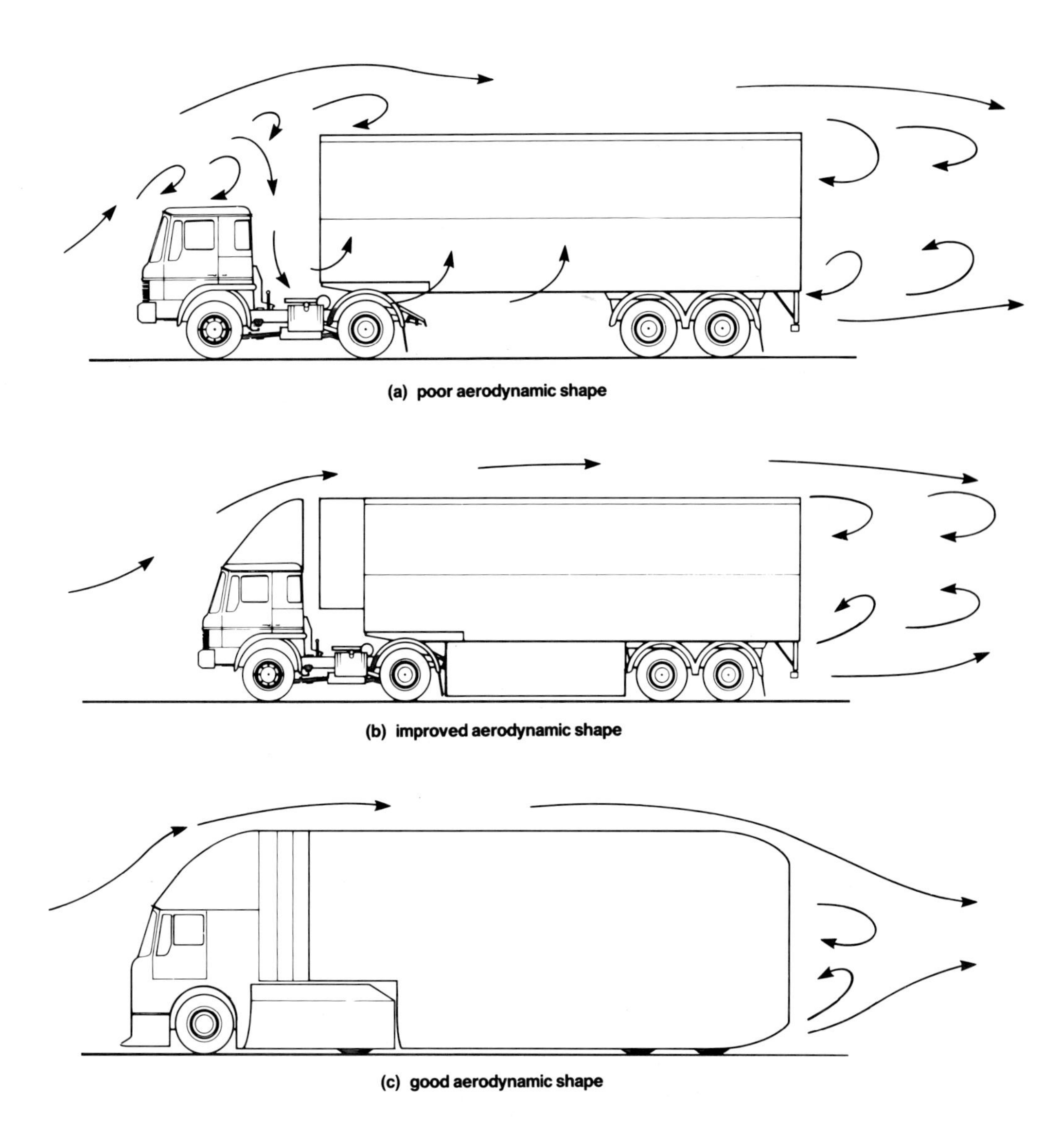

Fig 3.14 *Illustrations of improved aerodynamic shape for an articulated heavy goods vehicle (Source: Energy Efficiency Office, 1987)*

3.8 The effects of emission controls

Controls of exhaust emissions from road vehicles have a history connected with the special topography of the Los Angeles basin in California. By the mid-1950's, the high car use in Los Angeles was giving rise to the unpleasant photo-chemical smog, which reduced visibility only slightly, but caused irritation to eyes and throat, and accentuated breathing difficulties for people with asthma and bronchitis. The cause was the entrapment of cool air from the Pacific Ocean between the coast and the mountains to the east of Los Angeles. The cool layer of air did not allow car exhaust gases to disperse, and the action of strong sunlight on the oxides of nitrogen and unburnt hydrocarbons (fuel) produced unpleasant concentrations of ozone - the irritant gas.

Legislation, introduced in California in the late 1960's for the reduction of exhaust emissions, was made more stringent in later years and adopted by the US Federal Government. Emission controls in Europe and Japan followed the US lead, often with different control levels and different methods of testing - a feature which makes comparison as an international basis rather difficult. The progress of America emission control standards over the years is set out in Table 3.4. It may be mentioned in passing that the control of hydrocarbons is not restricted to exhaust emissions, but also now includes vapour from fuel tanks and fuel systems.

TABLE 3.4

US federal light duty emission regulation summary

Model Year	Exhaust emissions (g/mile)			Evaporation (g/test)	Particulate (g/mile)
	HC	*CO*	*NOx*		
Pre-control	15	90	6.2	6.0	-
1970	4.1	34	-	-	-
1971	4.1	34	-	6.0	-
1972	3.0	28	-	2.0	-
1973-74	3.0	28	3.1	2.0	-
1975-76	1.5	15	3.1	2.0	-
1977	1.5	15	2.0	2.0	-
1978-79	1.5	15	2.0	6.0(1)	-
1980	0.41	7.0	2.0	6.0	-
1981	0.41	3.4	1.0	2.0	-
1982-83	0.41	3.4	1.0	2.0	0.6
1987-88	0.41	3.4	1.0	2.0	0.2

(1) Change in test method: 6.0 by new method represents about 1.4 by the old.

Source: McArragh *et al*, 1989

These emission controls were introduced to improve air quality, and to protect public health. There is also (as has been seen in the Introduction) a concern over the increased emission of "greenhouse gases" which are now accepted as likely contributors to global warming. A brief description of the physics of global warming, and the relative importance of emissions from road transport, is in the Appendix A to this review. In the following paragraphs, the effects of emission controls, both for air quality improvement and for "greenhouse gas" reduction, are considered together.

The gases in the exhaust from a petrol or diesel engine contain the products of combustion of a complex hydrocarbon fuel and air. The main gases emitted are :-

Nitrogen (N_2)	originally in the air, and which has taken no active part in the combustion process.
Carbon dioxide (CO_2)	the product of complete combustion of the fuel.
Carbon monoxide (CO)	the product of incomplete combustion.
Hydrocarbons (HC)	unburnt or partially burnt fuel in vaporised form.
Oxides of nitrogen (NO_X)	mainly nitrogen oxide (NO) and small amounts of nitrogen dioxide (NO_2) resulting from high temperature reaction between atmospheric nitrogen and oxygen in the engine combustion chamber.
Water vapour (H_2O)	the product of complete combustion.
Particulates (especially from diesel engines)	very small particles of carbon from unburnt fuel. They may have polycyclic aromatic hydrocarbons and other complex organic compounds adsorbed on the surface.

Other gases may be present, including atmospheric oxygen which has not been used up in combustion, and possibly small quantities of nitrous oxide (N_2O) if a three-way catalyst is used (Warren Spring Laboratory, 1989).

The gases which are controlled for air quality reasons are CO, HC, NO_X: for diesel engines, particulate emissions are restricted. The gases which have most effect on global warming have been seen in Appendix A to be:-

Carbon dioxide (CO_2)	the major greenhouse gas.
Hydrocarbons and oxides of nitrogen (HC + NO_x)	because they react photochemically in the troposhere to produce ozone (O_3) which is a greenhouse gas.
Carbon monoxide (CO)	because it oxidises in the atmosphere to form more CO_2, and in doing so uses up hydroxyl radicals, and slows down the decay of methane (CH_4) produced from other sources. CH_4 is a powerful greenhouse gas.

Reduction of CO_2 emissions can only be made by reducing the quantity of fuel burnt, or by changing the fuel to one which contains less carbon*. By contrast, CO and HC emissions (and particulates) are the consequences of imperfect combustion processes within the engine and can,

---

* Proposals have been made for treating power station flue gases to remove CO_2 by liquification, but this is an energy intensive process and not practicable for road vehicles (See Baes, Jnr. *et al*, 1980).

in principle, be reduced by design changes based on research and development aimed at better understanding and improvement of the fuel burning process. Progress here will have the welcome effect of increasing combustion efficiency and thus reducing fuel consumption and CO_2 emissions. On the other hand, production of NO_X is a function of high combustion temperature in the engine, which is an indication of thermodynamic efficiency. So measures to reduce the formation of NO_X are unlikely to improve fuel consumption.

The way that CO_2, CO, NO_X, HC emissions and specific fuel consumption vary in two typical petrol engines as a function of air/fuel ratio is illustrated in Figure 3.15. For example, it can be

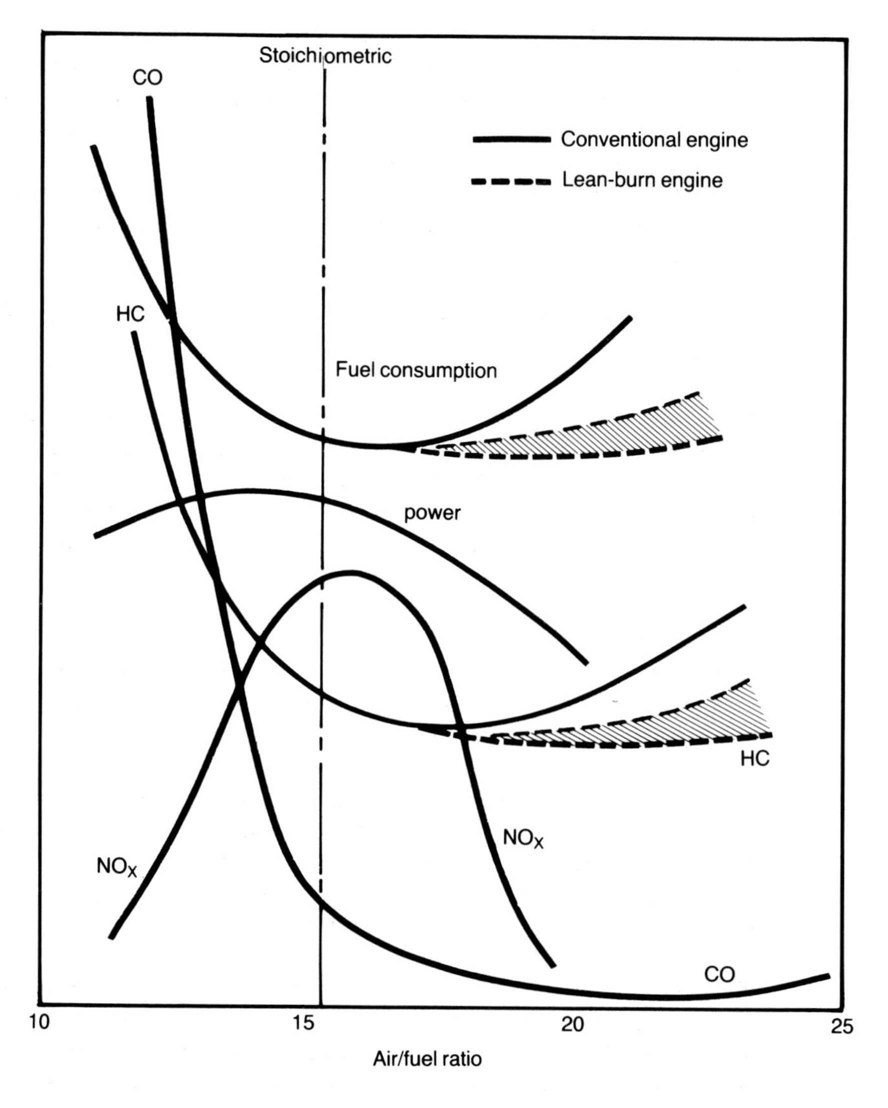

Fig 3.15 Relative relationships of typical engine emissions and performance to air/fuel ratio (Based on Blackmore and Thomas, 1977 and Felger, 1987)

TABLE 3.5
Effect of emission controls on petrol engine fuel consumption

Petrol Engine Technology	*Change in fuel consumption (%)*
Lean Burn	-14 to -1
Lean Burn + oxidation catalyst	-11 to 0
Lean Burn + oxidation catalyst + exhaust gas recirculation	-7 to +2
Three-way catalyst + oxygen sensor	+2 to +9
Non-optimised concepts	+3 to +12

Source: Martin and Shock, 1989, quoting Korte *et al*, 1987

seen that, for a lean-burn design, running at a high air/fuel ratio (eg. 18:1) reduces CO and NO_X and also fuel consumption and hence CO_2. A shaded band in the curves for fuel consumption and HC emissions indicates that, for some lean-burn engines, over-weakening of the mixture can lead to misfiring which reduces the advantage the design has over more conventional richer-burning engines. Much depends on the detailed design of the combustion chamber and the ignition control system if the potential advantages of lean-burn are to be attained in practice. It is useful to bear these graphs in mind when considering measures which are proposed and used to regulate emissions.

Various exhaust emission control technologies, and their effect on fuel consumption of petrol engines are listed in Table 3.5. The main options are:-

Lean burn engine - operating at high air/fuel ratios and with either carburettor or petrol injection fuel management. EGR (exhaust gas recirculation) into the cylinders is a means of reducing NO_X by inducing lower peak combustion temperatures, but with some fuel consumption penalty. Air injection into the exhaust manifold is a means of increasing the oxygen content of the exhaust gases, and so encouraging oxidation of CO and HC. Alternatively, an oxidation catalyst can be used to reduce the emissions of CO and HC.

Stoichiometric combustion - operating at the air/fuel ratio for complete combustion (about 15:1), but with a three-way catalyst which oxidises CO and HC, and reduces NO_X to nitrogen and oxygen. This "add-on" technology is proved and gives low emissions, but forgoes the improved fuel consumption that a lean burn engine could give.

An indication of the penalty (or benefit forgone) due to the use of catalysts has been obtained from a comparison of the official fuel consumption tests (Department of Transport, 1990) on pairs of car models which are identical apart from the fitment on one of a 3-way catalyst (Hickman and Waters, 1991). The comparison shows an average consumption 3.6% higher when a catalyst is fitted (with the range of results lying between an 11.5% increase and a 3.0% reduction).

If the most stringent emission controls are imposed (see Table 3.4), a three-way catalyst becomes the most likely technical solution for petrol engined cars. Low NO_X and particulate requirements may rule out diesel engines for cars, unless reliable particulate traps or oxidation catalysts can be developed. A fuller discussion of the technical and economic implications of regulations on emissions is given by Hickman and Mitchell (1990), and by Watkins (1991).

Emission controls are enforced by tests on new model cars, prepared by the manufacturer, and reinforced by very limited testing of new cars for "Conformity of Production" (this procedure is described by Fendick and Woolford, 1987). But there has been concern about how well the controls perform in the short term, between service intervals, and also over a long period, where catalyst performance (if one is used) may deteriorate. Taking the short term stability first, experimental work some years ago by Colwill *et al* (1985) showed how badly small cars with simple carburettors could get out of tune between service intervals, with consequent high levels of emissions (especially CO) and higher fuel consumption. Table 3.6 shows the results of tests on two small cars when new, and after one year's running. Carbon monoxide emissions have increased three-fold, and total hydrocarbons have more than doubled. The effect of service adjustments are shown to bring those emissions back to the "as new" state. Fuel economy would be expected to improve also, so that CO_2 emission would be reduced. These tests were on cars with carburettor fuel systems.

More recent trials with three similar cars with the same basic engine and modern engine management systems have given a more encouraging picture. One car had no extra emission control devices (non-catalyst or "uncontrolled"), one had a lean burn engine with oxidation catalyst, and the third had a three-way catalyst. The small change in average emission rates before and after service is shown in Table 3.7. With the non-catalyst car, HC and NO_X showed little

TABLE 3.6
ECE Regulation 15 tests on two cars at various intervals of time

	Emissions: Car A (g/test)			Emissions: Car C (g/test)		
	CO	*THC*	NO_X	*CO*	*THC*	NO_X
Post delivery inspection	92.8	12.0	7.7	55.2	9.8	7.6
First oil change	74.3	11.3	7.2	75.2	10.5	8.3
After 1 year						
- before service	275.9	25.4	1.6	280.5	24.1	0.9
- after service	52.4	9.4	6.7	62.8	9.6	7.9
Conformity of production standards	85	17.6	10.2	85	17.6	10.2

Source: Colwill *et al*, 1985

TABLE 3.7
Average emission rates (gm/km) for three modern cars before and after major services

Vehicle type	*Pollutant*	*Before*	*After*
Non-catalyst	NO_x	2.90	2.57
(Uncontrolled)	HC	0.70	0.81
	CO	4.60	6.25
Lean burn	NO_x	2.64	2.73
with oxidation	HC	0.12	0.13
catalyst	CO	0.08	0.09
Three	NO_x	0.46	0.41
way	HC	0.07	0.06
catalyst	CO	0.40	0.35

Source: Hickman, 1990

significant change after the service, while CO increased. For the other cars, there was no evidence of them going out of tune between service intervals.

The long term stability of the emissions of these modern cars is shown in Figure 3.16 by test results at constant speed taken at intervals during over 50,000 miles of normal driving. While there was a tendency for the controlled emissions (NO_X, HC and CO) to increase with miles

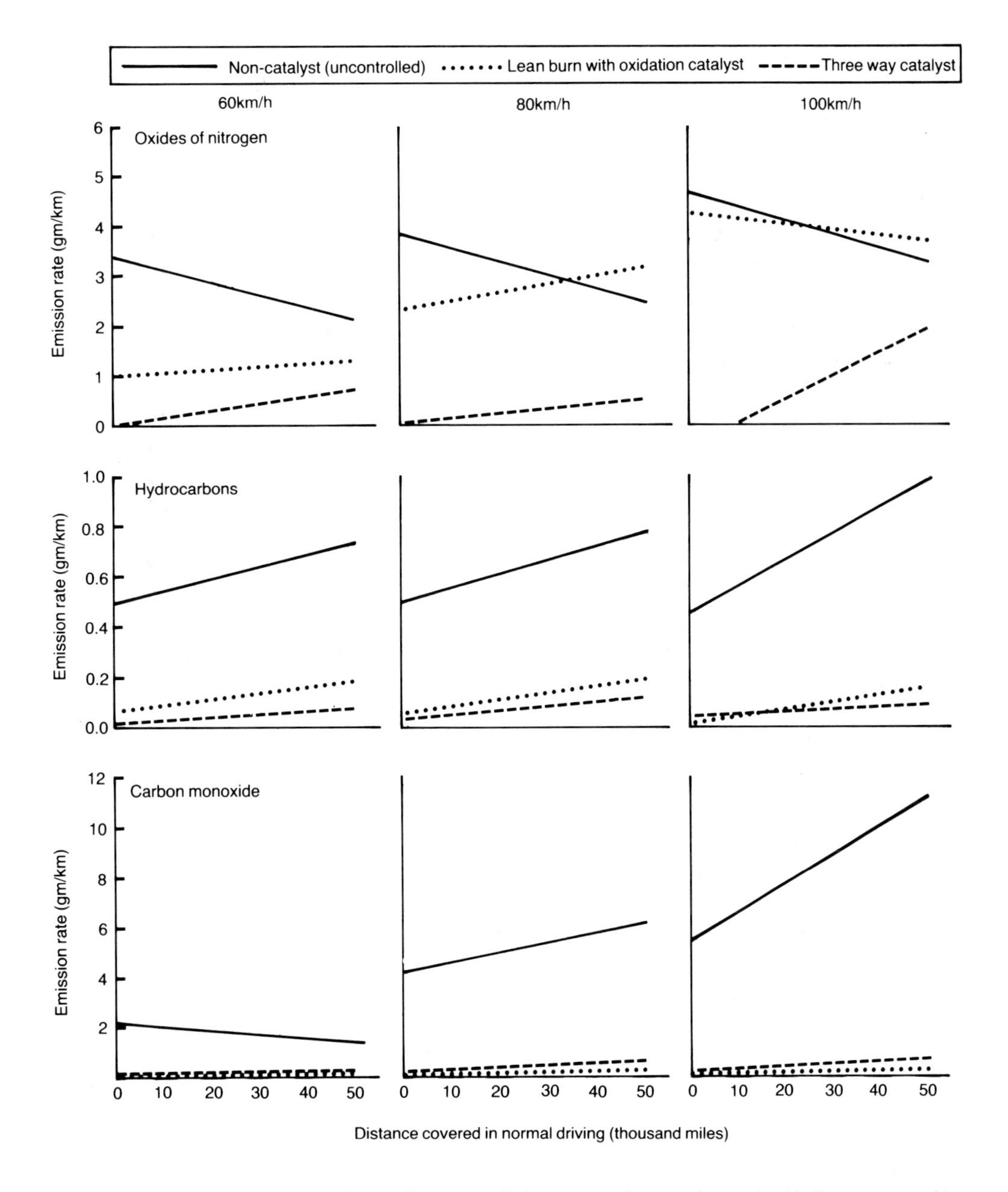

Fig 3.16 Three modern cars: variation of exhaust emission rates at three steady speeds with distance covered in normal driving (Source: Hickman, 1990)

covered in service, emission levels from the lean burn and three-way catalyst cars remained much lower than the non-catalyst car, except for increase of NO_X, where the non-catalyst car had slightly lower emissions than the lean burn car at the two higher speeds after about 30,000 miles of normal driving had been completed.

The change in fuel consumption with miles driven is shown in Figure 3.17. All cars had a tendency for the fuel consumption to increase with mileage, with the uncontrolled car having the greatest change. (A cyclic variation apparent in the results may be due to seasonal changes in ambient conditions and fuel quality.) The lean burn car had the lowest consumption, but all three types were notably economical when compared with the 1986 fleet average fuel consumption of nearly 10 litres/100 km. The absolute values of fuel consumption must be treated with caution because the three cars had somewhat different road performances, with the "uncontrolled" car, for example, being a 'sports' version of the model.

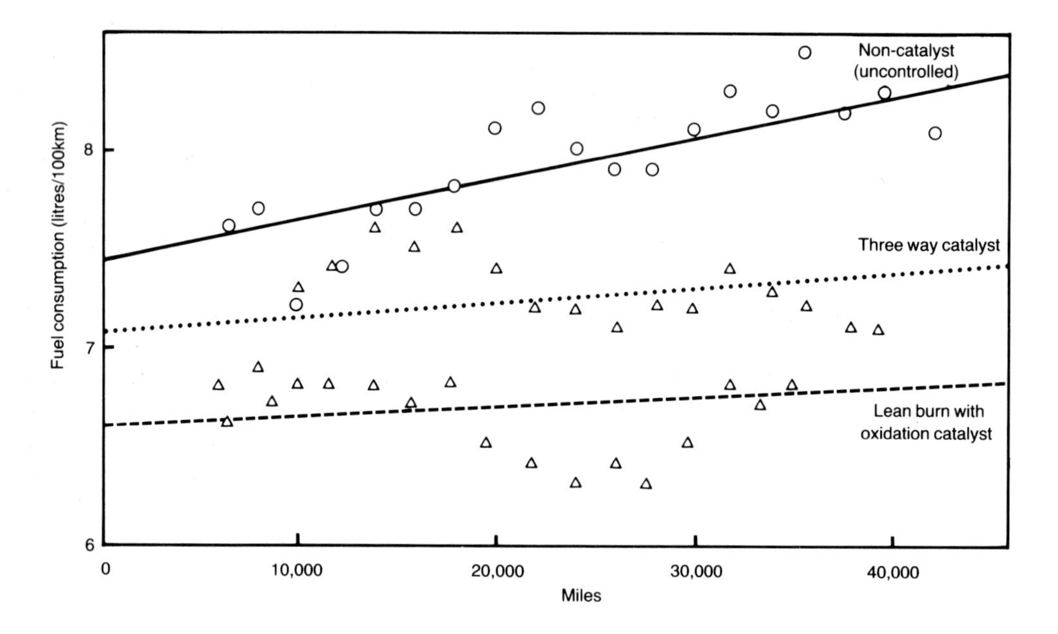

Fig 3.17 Three modern cars: variation of fuel consumption with distance covered in normal driving (Source: Hickman and Pearce, 1988)

The main objective of emission controls is to improve air quality, especially around roads carrying heavy traffic. There is, therefore, an interest in predicting the average emissions from a stream of traffic, and one way of doing this is by the development of "emission factors" which represent the emissions of groups of similar vehicles under similar driving conditions. The methodology used in one study for the European Commission has been described by Eggleston *et al* (1989). The emission factors can be aggregated to represent the features of different traffic flows under urban or rural conditions by using average speed as a variable. As a by-product of the method, information also becomes available on average fuel consumption - because analysis of the exhaust gases gives the total carbon content, and enables fuel use to be calculated. But the data has limitations when applied to individual car designs, and there are also shortcomings in basic information which is available so that it is difficult to include all the variables which are known to be relevant to emissions and fuel consumption.

Further information on the effects of journey types and car speeds (and driver characteristics - see Chapter 4) is being obtained by a large international study which is in progress under the EC DRIVE programme. Participants are TRRL (UK), INRETS (France), CEDIA (Belgium) and TUV Rheinland (Germany). The first stage of measuring driver behaviour and vehicle operations in actual use in urban areas is being planned and carried out under the acronym "MODEM", which stands for "modelling of emissions and consumption in urban areas" (Andre *et al*, 1989).

Eventually this research will help to improve the prediction of air quality benefits resulting from changes in vehicle design to meet new emission standards, and should also illuminate any consequent changes in fuel consumption from the emission control systems used.

This short review of the effects of emission controls on fuel consumption has been mainly about petrol engines for cars. For the large diesels used in HGVs, a review of the implications for engine design of various stages of regulation for reduced emissions was carried out for TRRL by Ricardo Consulting Engineers plc (Latham *et al*, 1988). Seven levels of increasing severity of control were considered, concentrating on HC, NO_X, and particulates. They are shown graphically in Figure 3.18. Level 1 corresponds to the ECE Regulation 24; Level 3 to EC 1988/90 regulations; Level 7 to the proposed 1994 US Requirements. (Approximately equivalent levels to the last two are given in Table 3.8 in terms of grams/kWh). It was concluded that all current heavy duty diesel engines could meet present and proposed European standards for gaseous emissions with little or no adjustment. To achieve further reduction in smoke or particulates, technically feasible, but unproven, technologies would be needed. The most severe emission controls (Level 7) would pose considerable problems to the engine manufacturers. An initial increase in price would be expected, and improvements in fuel economy, possible for markets without emission controls, would not be achieved. It seems that advanced fuel injectors and turbocharging/aftercooling engines will be needed, and particulate traps (not yet developed) will have to be fitted. Latham *et al* suggest that these penalties will stimulate interest in other fuel economy measures like advanced transmissions and improved aerodynamic shape to assist in retaining good fuel consumption.

TABLE 3.8
Emission legislation for heavy duty diesel engines

Emission product	*EC 1988/1990 g/kWh*	*US Federal 1994 g/kWh**
HC	2.4	1.7
CO	11.2	20.8
NO_x	14.4	6.7
Particulates	Not yet defined	0.1

*Converted from g/BHPh from the US test cycle
(Source: Martin and Shock, 1989)

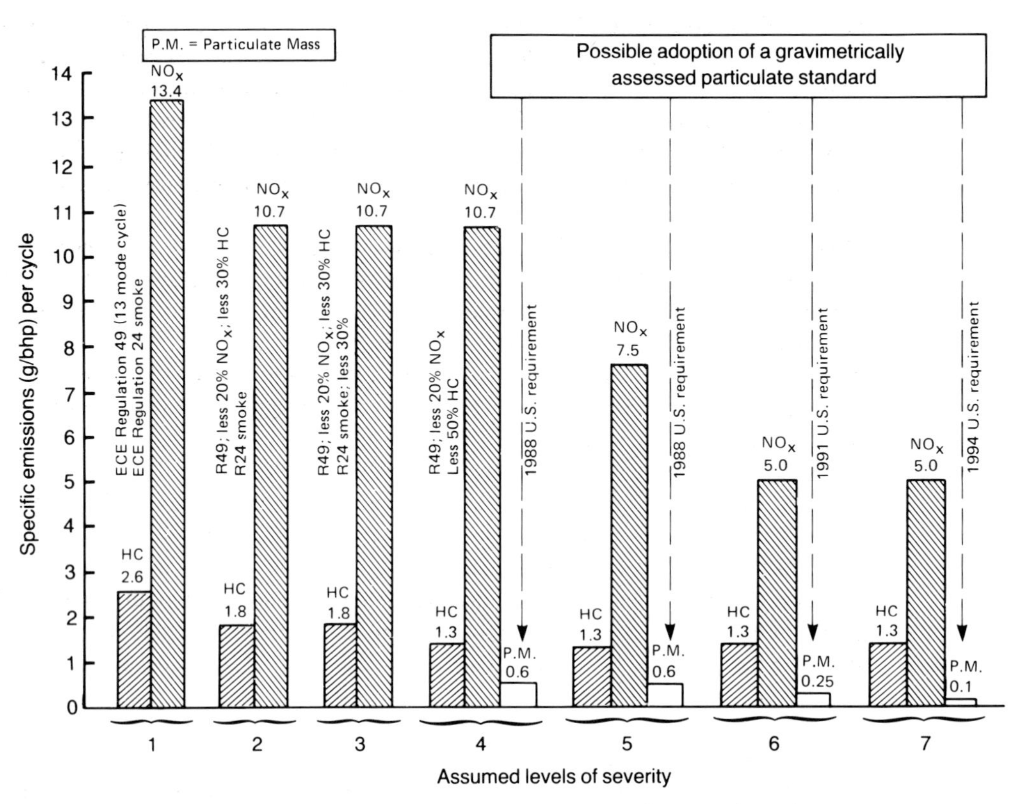

Fig 3.18 *Possible diesel exhaust emission regulation levels (Source: Latham and Tonkin, 1988)*

However, a more recent assessment by the Department of Transport is that the US 1994 standards should be achievable without the use of particulate traps. The technology needed includes electronic fuel injection and possibly oxidation catalysts which will become available in the next few years. For this reason, it is now UK Government policy to support the introduction of EC Standards equivalent to the US 1994 standards, to come into effect in 1995/96 (CMEO, 1990).

3.9 Future possibilities for the improvement of fuel economy

The different design factors, reducing fuel consumption, or exhaust emissions, or both, can be combined in many ways to give estimates for different future years. For example, one early "target" for an economy car suggested the following potential savings, based on the application of proved technology :-

Reduced weight, drag and rolling resistance	10%
Matching of engine and transmission for economy	10%
Change to high efficiency engine (diesel)	25%
Reduction of out-of-tune conditions	5%

Overall, these possibilities lead to an economy car with a fuel consumption in mixed urban and rural driving of 6 litres/100km, (compared with a base value of 10) (Waters and Laker, 1980). The development of "concept cars" by manufacturers has shown that this modest target is achievable, and can be bettered. The British Leyland ECV 3, an economy concept vehicle, (King, 1984) gave a fuel consumption of 4.5 litres/100km based on a mix of urban, rural and motorway driving, though it is not clear that it would meet the tighter emission regulations of the 1990s.

However, the Volvo LCP 2000 concept car is a newer design, a two seater with a large luggage compartment or the ability to carry two extra (rear-facing) passengers. It has a small 3 cylinder direct injection turbo-charged diesel engine, and meets the US 1983+ emission rules (Table 3.4), except for HC, where it is 5% over the limit. In the same mix of driving it achieves a fuel consumption of 3.6 litres/100km (Volvo, 1990). Mellde *et al* (1989) quote the VW Auto 2000 diesel car as attaining 3.6 litres/100km in mixed city and highway driving, and the Toyota AXV diesel as returning 2.4.litres/100km.

Another recent demonstration car, the VW Futura, uses a supercharged direct injection petrol engine based on the Golf diesel and operating at a compression ratio of 16:1, and at air/fuel ratios up to 80:1 at part load. A fuel consumption of 5.7 litres/100km. is claimed with the engine in a standard Golf body. This is only 2% greater than the consumption of the latest Golf diesel car (Autocar and Motor, 1990). These examples, although they are not production cars, go to show that the potential for technical improvements to the fuel consumption of the private car is still substantial.

A more comprehensive look at technical factors and fuel consumption for cars was carried out by Francis and Woollacott (1981), but, again, emission requirements have become more difficult to meet since then. The most recent set of future projections of fuel consumption, for the year 2010 is that made by Martin and Shock (1989). Table 3.9 summarises their judgements made for different sizes of car, and for different classes of road. It should be noted that the recent agreements in the European Commission on emissions mean that the last entry in the Table "overall - with stringent emission regulations" must now be expected to be the norm. The loss of possible fuel consumption improvement for the smaller cars is evident. This is because the economical lean-burn engine can no longer be used, and it is expected that three-way catalysts will be needed for all car sizes. The resulting average fuel consumption is then estimated for the private and business motoring fleet in the year 2010, and shown in Table 3.10. Quite modest improvements are evident, compared with what has been seen to be technically possible, and reflect the estimate made of the take-up of technology in the car fleet as driven by expected market forces, and the effects on fuel economy of stringent controls on exhaust emission. The small improvement for diesel cars shows the likely effect of controls in NO_x and particulates.

From the same report, the forecast changes in fuel consumption for diesel engined goods vehicles is shown in Table 3.11. As expected, the reductions in fuel consumption are less than for cars, and if the US Federal 1994 emission regulations (Table 3.8) are applied, fuel consumption could increase. The projected average fuel consumption for the year 2010 is given in Table 3.12 for the different classes of goods vehicle. It can be seen that for all classes some increase in fuel consumption is predicted. To complete the picture for goods vehicles, Martin and Shock give their estimate of the likely average fuel consumption for light and medium weight vans. They are shown in Table 3.13, where it can be seen that with stringent emission standards, fuel

TABLE 3.9
Changes in petrol car fuel consumption by rapid technical change by the year 2010

Item	*Feature*	*Engine size*	*Built-up roads*	*Non built-up roads*	*Motorways*
I	Emissions control technology to satisfy currently proposed regulations	up to 1.4 l	-12%	-10%	-8%
		1.4-2 l	-12%	-10%	-8%
		> 2 l	+3%	+3%	+3%
II	Development of engine technology (such as reduced friction, precision cooling, reduced pumping losses)	up to 1.4 l	-5%	-4%	-3%
		1.4-2 l	-5%	-4%	-3%
		> 2 l	-5%	-4%	-3%
III	Transmission developments (eg automated manual)	up to 1.4 l	-8%	-4%	0
		1.4-2 l	-8%	-4%	0
		> 2 l	-8%	-4%	0
IV	Vehicle weight reduction by 10 per cent	up to 1.4 l	-5%	-2%	0
		1.4-2 l	-5%	-2%	0
		> 2 l	-5%	-2%	0
V	Aerodynamic drag/rolling resistance reduction by 10 per cent	up to 1.4 l	-2%	-4%	-6%
		1.4-2 l	-2%	-4%	-6%
		> 2 l	-2%	-4%	-6%
	OVERALL - WITH CURRENT EMISSIONS REGULATIONS (ie ITEMS I, II, III, IV & V)	up to 1.4 l	-28%	-22%	-16%
		1.4-2 l	-28%	-22%	-16%
		> 2 l	-16%	-11%	-6%
VI	Emissions control technology to satisfy stringent regulations	up to 1.4 l	+3%	+3%	+3%
		1.4-2 l	+3%	+3%	+3%
		> 2 l	+5%	+5%	+5%
	OVERALL - WITH STRINGENT EMISSIONS REGULATIONS (ie ITEMS II,III,IV,V & VI)	up to 1.4 l	-16%	-11%	-6%
		1.4-2 l	-16%	-11%	-6%
		> 2 l	-14%	-9%	-4%

Notes: '+' values indicate an increase in fuel consumption
'-' values indicate a reduction in fuel consumption

Effects are expressed as percentage improvements over 1986 average, and assume that the current R & D potential in vehicle engineering is realised by rapid technical change by the year 2010.

The "stringent" regulations are now those currently proposed.

Source: Martin and Shock, 1989

TABLE 3.10
Estimated average fuel consumption for cars in 2010

Vehicle	*Less stringent emission control (Rapid technical change)*	*Stringent emission controls*	*1986 Reference*
Petrol engined cars	7.4	8.4	9.6
Diesel engined cars	5.4	5.8	5.9

Fuel consumption in litres/100km

Source: Martin and Shock, 1989

TABLE 3.11
Changes in diesel vehicle fuel consumption by rapid technical change by the year 2010

Item	Feature	Vehicle Size	Built-up roads	Non built-up roads	Motorways
I	Emissions control technology to satisfy proposed regulations	Light goods	+1%	+1%	+1%
		Medium trucks	0	0	0
		Heavy trucks	0	0	0
II	Developments of engine technology (such as unit injectors, reduced friction)	Light goods	-4%	-4%	-4%
		Medium trucks	-4%	-4%	-4%
		Heavy trucks	-4%	-4%	-4%
III	Transmission and powertrain developments (such as turbo-) compounding, integrated control	Light goods	-4%	-2%	0
		Medium trucks	-4%	0	0
		Heavy trucks	-2%	-1%	-4%
IV	Aerodynamic drag/rolling resistance reduction by 20 per cent	Light goods	-2%	-5%	-8%
		Medium trucks	-2%	-6%	-10%
		Heavy trucks	-2%	-6%	-10%
	OVERALL - WITH CURRENT EMISSIONS REGULATIONS (ie items I, II, III, & IV)	Light goods	-9%	-10%	-11%
		Medium trucks	-10%	-10%	-11%
		Heavy trucks	-8%	-10%	-14%
	MONITORING AND TARGETING	All vehicle types	-5%	-5%	-5%
V	Emissions control technology to satisfy stringent regulations	Light goods	+20%	+20%	+20%
		Medium trucks	+15%	+15%	+15%
		Heavy trucks	+15%	+15%	+15%
	OVERALL - WITH STRINGENT EMISSIONS REGULATIONS (ie items II, III, IV & V)	Light goods	+8%	+7%	+6%
		Medium trucks	+4%	+4%	-1%
		Heavy trucks	+6%	+5%	-1%

Notes: '+' values indicate an increase in fuel consumption
'-' values indicate a reduction in fuel consumption

Effects are expressed as percentage improvements over 1986 average, and assume that the current R&D potential in vehicle engineering is realised by rapid technical change by the year 2010.

The "stringent" regulations are now under discussion.

Source: Martin and Shock, 1989

consumption is expected to increase except for petrol fuelled car-derived vans and light utility vehicles.

Martin and Shock's conclusions on the effects of the US 1994 emission controls on fuel consumption of goods vehicles are now regarded as over pessimistic. The application of electronic fuel injection and oxidation catalysts, mentioned earlier, are now thought to enable the lower emissions to be reached with a fuel consumption penalty of nearer 5% than 15%-20%. Advances in aerodynamics and non-engine factors should then provide a net improvement in vehicle fuel economy by the year 2010, assuming no further tightening of NO_x and particulate limits beyond the US 1994 levels. This view is reinforced by a recent analysis of trends in HGV fuel economy, which has shown, in broad terms, an improvement of between 0.5% and 1.0% per year over the past decade (CMEO, 1990).

TABLE 3.12
HGV projected average fuel consumption by the year 2010

Vehicle Category		Average fuel consumption (litres/100 km) *Less stringent emission controls (Rapid technical change)*	*Stringent emission controls*	*1986 Reference*
Heavy Vans		13.7	16.1	15.2
2 axle	up to 7.5t	15.9	18.4	17.7
	7.5 to 12t	17.7	21.6	20.8
	12 to 15t	18.7	22.9	22.0
	>15t	19.9	24.3	23.4
3 axle rigid		24.5	30.0	28.8
4 axle rigid		27.7	34.2	32.6
Articulated	up to 25t	28.4	34.7	33.4
	25 to 32.5t	32.0	39.5	37.6
	>32.5t	34.8	43.0	40.9

Source: Martin and Shock, 1989

TABLE 3.13
Light goods vehicle projected average fuel consumption by the year 2010

Vehicle Category	Average fuel consumption (litres/100 km) *Less stringent emission controls (Rapid technical change)*	*Stringent emission controls*	*1986 Reference*
Car derived vans			
- petrol	7.8	8.8	9.7
- diesel	6.3	7.6	7.0
Light utility vehicles	8.9	9.2	10.5
Medium vans			
- petrol	14.5	16.7	15.9
- diesel	10.5	12.6	11.7

Source: Martin and Shock, 1989

3.10 Summary

The general conclusions on design factors are that technical possibilities seem likely to be considerably modified, and the fuel economy improvement reduced, by the combined effects of more exacting emission regulations, and the rate of take-up of technology in the fleet. This latter is, in essence, a judgement about the market and other forces pushing vehicle design towards lower fuel consumption. The consumption of diesel powered goods vehicles seems likely to stay static or at best improve only slightly if (when) the rigorous controls of NO_x and particulates are introduced. For cars, however, the most recent concept cars, usually as small as practicable, have demonstrated in prototype form that overall fuel consumption of around 3 to 4 litres/100km is possible in small vehicles. The present generation of "super-mini" petrol cars, now in production, have a fuel consumption, based on regulatory tests, of between 5 and 6 litres/100km, which is 25% less than the average for new petrol cars. New technology and smaller cars appear to be major factors in improving fuel efficiency. The challenge is to make this performance the average for the whole fleet, while maintaining and improving the protection given to occupants in an accident.

4 Driver, traffic and fuel economy

4.1 Factors affecting fuel economy

In the previous Chapter, the influence of engine and vehicle design on fuel economy was the main consideration, although the effect of driving the vehicles on different kinds of road was implicit in some of the figures quoted. In this Chapter, the emphasis changes to how the vehicle is driven by real drivers, in real traffic on real roads, and what factors determine the fuel economy that is (or can be) achieved. The kind of questions that will be addressed include:-

- How does the presence of other traffic affect fuel consumption?
- What role can traffic management and speed limits play?
- How do drivers react to traffic conditions?
- What information and training can be useful to drivers?
- If drivers fit "add-on" fuel economy devices to their vehicles, can they save fuel?

Most of the following discussion, but not all, will be about cars and car drivers. One reason is that much of the information available from research and trials is about cars, but another is that the way the driver drives his car in traffic has much more effect on fuel consumption than the way the lorry driver drives his lorry. This is because the much higher power/weight ratio of the car* allows, or even encourages, the car driver to drive in an uneconomical fashion. However, where information exists about goods vehicles, it will be used, and there will be some discussion of the role that driver motivation and training can play in commercial vehicle operation.

4.2 The effect of traffic

There is a significant difference between the fuel consumption of an isolated car, driven at various steady speeds, and the consumption of the same car travelling at the same average speed but constrained to follow the start/stop progress of a typical traffic stream. (This was evident in the illustrations of the Golf cars in Figures 3.1 and 3.2). For the isolated car, fuel consumption reduces with steady speed, usually down to a speed at which it is no longer comfortable to stay in top gear. There is a minimum fuel consumption, but it is at a speed (30 - 40 kph) which is too low to be of serious interest to most motorists.

With a car in a stream of traffic, the situation is quite different. Fuel consumption falls as speed is reduced from motorway speeds, but there is a distinct minimum at around 60 km/h before the inefficient stop/start cycle of congestion takes over, and fuel consumption rises. This relationship

* Typically, a car has a power/weight ratio of about 40 kW/tonne compared with that of a loaded H G V of around 5 kW/tonne.

was widely publicised by Everall (1968) who followed up earlier work at RRL (for example, by Lister and Kemp, 1954, and others). Everall's research was extensive, and covered cars, light and heavy goods vehicles, the effect of different drivers, and of gradients. The objective was to provide data for vehicle operating costs to be used in cost/benefit assessments of road schemes. The classic "Everall curve" of fuel consumption as a function of section average car speed is reproduced in Figure 4.1, and regression equations were developed to quantify fuel consumption as a function of speed. The form of the general relationship was:-

$$C = A + B/V + C.V^2$$

where C = fuel consumption (eg litres/100 km)

V = average vehicle speed over the section of road.

A, B, C are constants

This work on traffic-influenced fuel consumption was extended to include other road design variables by Langdon (1984), who concluded that speed terms alone only accounted for about 35% of the variability of the new data. Additional terms relating to gradients and road design features (like roundabouts), which caused vehicles to accelerate or decelerate, gave a more satisfactory equation. Langdon's results are particularly useful when considering the effects of changing road layouts, by realignment or the construction of by-passes. Watson *et al* (1980) also improved the explanatory fit of the speed term regression equation: he proposed including a parameter, PKE, defined as the sum of the positive kinetic energy changes in a trip. With this parameter, 88% of the observed variability in urban journeys could be explained with a simple four term equation (ie: constant; V [vehicle speed]; 1/V; and PKE).

A description of a hierarchy of vehicle fuel consumption models has been given by Akcelik *et al* (1983). They range from detailed models for automotive engineering purposes (level O models), through to coarser models for traffic, transport and urban form analysis (level I to level III models). For example, Watson's PKE model (above) is treated as a level II model, useful for traffic and transport management purposes. The valid point being made is the need to match the type (and complexity) of model with the use intended for it.

Although more complex models exist, the Everall relation, depending only on speed terms, has proved a valuable means of analysing experimental results, especially in urban areas. It was used in an assessment of the effect of traffic congestion on fuel consumption (Gyenes, 1980), where the aim was to identify that part of UK road transport fuel consumption which was attributable to traffic congestion, and could possibly be reduced by measures which modified traffic flow. The results were rather surprising in that reducing congestion from peak time levels to off-peak levels only appeared to save 1.5% of total fuel used. If peak congestion could be reduced to ideal free-flow speeds of around 50km/h, the potential saving of fuel rose to 6.1%. One reason given for the comparatively small savings was that only about 11% to 13% of total fuel was used in the congested central areas of towns and conurbations. But as the results were based on 1976 traffic data, it might be expected that present traffic conditions would represent a larger fuel penalty: a re-calculation for current conditions would be interesting.

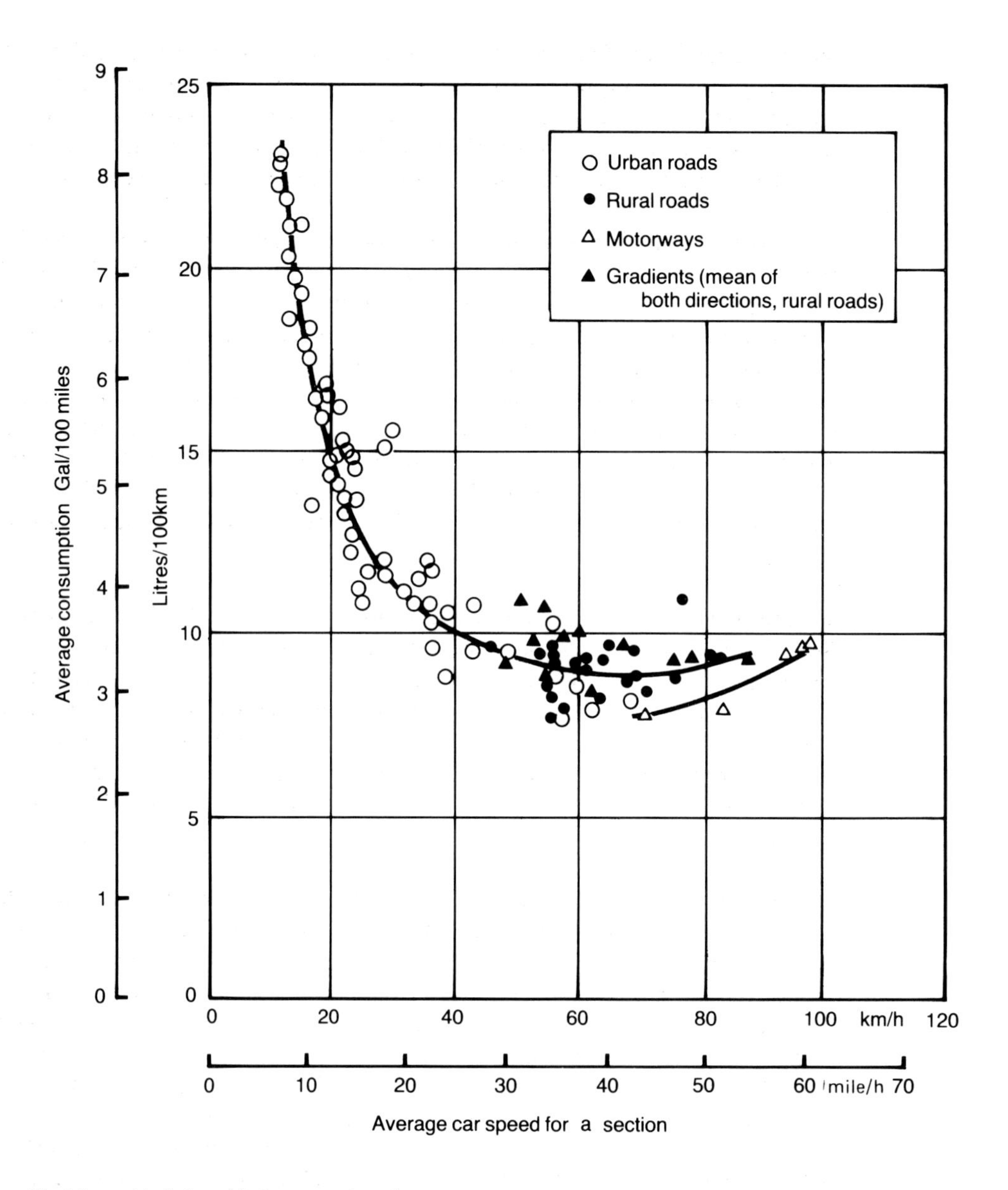

Fig 4.1 *Variation of fuel consumption of an average car (Source: Everall, 1968)*

4.3 The road system

The increased fuel consumption in congested traffic suggests that providing sufficient capacity in the road system must be a high priority in reducing fuel use. The question of whether increasing road capacity causes an increase in traffic, (and hence, in total, CO_2 emissions) has been the subject of much debate and is beyond the scope of this review. In any case, the Secretary of State for Transport commented in 1990 that it will not be practical to increase capacity to match the forecast increase of traffic. It is not possible to go into arguments for and against traffic restraint in this paper: it is merely noted that the design of the road system has a great influence on vehicle fuel consumption and emissions. In this Section, some of the effects of traffic management schemes, speed limits, and road surface roughness are examined briefly. The possible benefit, in fuel consumption terms, of route guidance for the driver is also touched on.

The beneficial effects of smoothing traffic flow in urban areas has been mentioned before. Watson *et al* (1980) with the PKE term, and Langdon's (1984) vehicle acceleration/deceleration factor are evidence that greater smoothness in the traffic reduces fuel consumption. This idea was applied to area traffic signal systems, notably by Robertson *et al* (1980), who showed how the TRANSYT computer program could predict the fuel consumption in a traffic network, and then be used not only to minimise vehicle delay, but additionally to minimise the number of starts and stops. An extra saving of about 3% of overall fuel was achieved in real traffic by the last step. These savings have been confirmed by more recent work, reported by Mulroy (1989), where a new computer program developed by TRRL (SCOOT) gave savings of 5% more fuel in the morning peak period than the fixed time Urban Traffic Control system (TRANSYT) when applied in the town of Worcester.

The effect of other changes in traffic management were studied by Wood and Griffin (1980) in Swindon, where traffic lights and uncontrolled junctions were replaced by mini-roundabouts. It appeared (in spite of experimental difficulties) that fuel saving in the range 1% to 4% was obtained by the traffic management measures. A specific study of fuel consumption (Gardiner *et al*, 1986) has led to the development of simple equations for estimating the fuel use due to roundabout geometry, and the consumption in traffic queues controlled by a priority system. The method enables a more accurate account to be made of the costs and benefits of different roundabout designs.

Turning from urban roads to roads outside built up areas, there has been considerable discussion on the value of speed limits in reducing fuel consumption, and action has been taken in several countries. In the UK, a 50 mph speed limit on motorways and other non-built-up roads was imposed during the first oil supply crisis in 1973/74. In the USA, a 55 mph limit has been in place in most States since that time, though it has recently been relaxed in some. Sample calculations made in 1974 suggested that, in the UK, compliance with a 50 mph speed limit might save around 2% to 3% of car fuel* (1.5% to 2% of road transport fuel) (Waters, 1977). Another study by Leake (1980) also concluded that only about 1.5% of car fuel would be saved by a 50 mph limit. Estimates from American work indicated that savings from cars were unlikely to be greater than 2.5% of total car fuel, even for 100% compliance with a 55 mph limit (Waters and Laker, 1980).

Compliance with a speed limit is, of course, crucial. The UK experience during the 50 mph limit period is interesting. On 19 November 1973, drivers were requested voluntarily not to exceed 50 mph on any roads, including motorways, and appeals were made for unnecessary journeys to be avoided. Exhortations to drive "more gently" were made, and these appeals were given force because of a shortage of petrol at many filling stations in London and the South East. On 8th December 1973, the 50 mph limit was made compulsory on all roads not subject already to a

---

* An unpublished re-calculation with 1982 traffic and vehicle characteristics suggested that the car fuel saved might have increased to about 4%.

lower limit. This mandatory limit lasted until 29 March, when the 70 mph limit was restored on motorways. On the 8th May 1974, the 60 mph limit was reinstated on all-purpose roads.

Between November 1973 and November 1974, the retail price of petrol had increased from about 35p/gallon to about 63p/gallon.

One result of this complex series of events is shown in Table 4.1, which gives measurements of the speeds of cars on two motorways in the South East. The voluntary 50 mph limit on mid-November 1973, together with physical shortage of fuel reduced the mean speed from 70 mph (with 49% exceeding 70 mph) to 58mph (with only 10% exceeding 70 mph). The mandatory 50 mph limit reduced the mean speed to 54 mph, with only 1% of cars still exceeding 70 mph. But before the limit was rescinded (in early May 1974), mean car speeds had increased to 66 mph and 34% were exceeding 70 mph. It may be significant that supplies of petrol, even at a higher price, were now plentiful.

TABLE 4.1
Speeds of cars measured on M3 and M4

Date of measurements		*Mean speed (miles/h)*	*% exceeding 70 mile/h*
August-October	1973	70	49
November 21-22	1973	58	10
December 4-6	1973	56	9
December 12-13	1973	54	1
January	1974	57	3
February	1974	60	13
April	1974	66	34
September	1974	70	50

Source: Scott and Barton, 1976

Much the same picture emerges from Figure 4.2 where measurements of traffic speed on the M4 motorway, and lead concentration in the air (a measure of fuel used) are plotted over the same time period. It is clear that initial compliance with the 50 mph limit soon eroded, though it is not so obvious why lead levels remained at a lower level than before the speed limits were imposed. However, there were other benefits in reduced accidents over the whole period which have been evaluated by Scott and Barton (1976).

It is evident that lower speed limits are not likely to be a major source of fuel saving in the UK, but new estimates with higher traffic flows on motorways and other high speed roads may be needed to give a more up to date value of savings that might be expected.

Similar calculations were carried out by Roumegoux (1983) for conditions in France. In urban areas, it was concluded that the effect of driving style on fuel consumption was so great that it could not be certain that fuel would be saved if a 50 km/h limit were introduced. But if the present 60 km/h limit was observed by drivers, large savings would result (45,000 tonnes). Larger savings of petroleum would accrue if speed limits on inter-urban roads were observed (300,000 tonnes) and even more if the speed limits on these roads were reduced by 10 km/h (saving a further 200,000 tonnes). Unfortunately, Roumegoux did not quote the total petroleum consumption, so

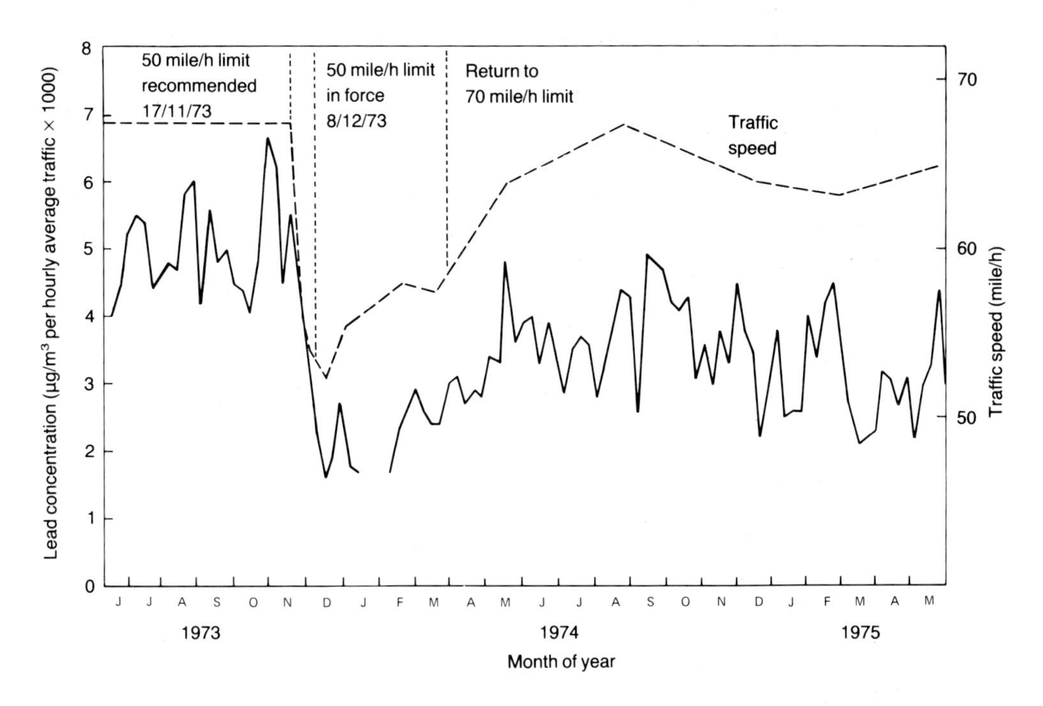

Fig 4.2 *Traffic speed and lead concentration (as a proxy for fuel consumption) on the M4 motorway, 1973-74 (Source: Hogbin and Bevan, 1976)*

that it was not possible to put these savings directly in perspective. However, from another source (CEC, 1988) the total energy consumption for road transport in 1983 in France is given as 28.9 MTOE. The maximum savings on urban and inter-urban roads (545,000 tonnes) thus represents about 2% of road transport fuel - a result not inconsistent with the UK estimates for rather different conditions.

Strict enforcement of heavy goods vehicle speed limits could produce more substantial savings per vehicle, even though in total they use only about 20% of road transport fuel. There are certainly worthwhile savings of fuel cost to be made by the operator. Shell UK have said that limiting the top speed of a commercial vehicle could be the most significant factor in a fuel saving programme. Research has shown that for an HGV at 32.5 tonnes GVW, increasing speed from 100 to 110 km/h can increase fuel consumption by 14%: reducing speed to 90 km/h can produce a fuel saving of about 12% (quoted in Energy Efficiency Office, 1987). As the heavier goods vehicles have a high proportion of mileage on motorways, the effect of speed restrictions will be more marked than with cars*. Table 4.2 shows that articulated heavy goods vehicles run 45% of their mileage on motorways, compared with 14% for cars. (For running on both motorways and

*In May 1991, the UK Minister for Roads and Traffic announced that draft regulations would require all new HGV's over 7.5 tonnes GVW to be fitted with speed limiters after 1st August 1992. He also proposed to bring forward regulations requiring speed limiters to be fitted to articulated vehicles first registered after 1st August 1988. (Department of Transport, 1991)

TABLE 4.2
Car and HGV traffic on different classes of road: 1989*
(Billion vehicle kilometres, and %)

Road Type	Cars/Taxis	All HGVs	of which: Artics	All Motor Vehicles
Motorways	44.8	9.1	4.6	59.5
Major roads (non-built up)	89.9	10.4	3.9	112.1
All roads	327.0	29.7	10.1	402.6
Motorways/All	13.7%	30.5%	45.0%	14.8%
Motorways + Maj/All	41.2%	65.3%	83.2%	42.6%

* Provisional
Source: Department of Transport, 1990b

non-built-up roads, the corresponding figures are 83% and 41%.) Thus speed restrictors would have a larger proportionate effect for HGV's than for cars, though since cars account for a much larger fraction of the total fuel used the absolute fuel savings from car speed restrictions could be substantial.

Another factor which has an influence on vehicle fuel consumption is the roughness of the road's running surface. A rough road will generate vertical motion which causes tyre and suspension deflection to absorb the energy. The energy loss appears as increased fuel consumption. Measurements with cars and lorries have provided the averaged fuel/unevenness relationships shown in Figure 4.3. For the range of unevenness likely to be found on major roads in the UK, the increase in fuel consumption might be expected to be in the range 3% to 4%. This is in broad agreement with similar work carried out in France. It was concluded that the effect of road unevenness might be more significant for lorries than for cars, though further work is needed to confirm this. The total cost of increased fuel consumption is likely to be greatest on motorways, in spite of their usually good surface, because of their high traffic flows.

4.4 Driver characteristics in traffic

It is well known that different drivers can obtain substantially differing fuel consumption figures in the same model of car. For example, fuel economy (miles per gallon) figures 50% greater than those obtained by the motoring press drivers were achieved by expert drivers in a well known fuel economy competition for a wide range of car models (Waters and Laker, 1980). But these kinds of result are from uncontrolled tests, and are not representative of the variations between average drivers under repeatable conditions. Controlled trials can be divided into those where drivers drive over a set test route, in the absence of other traffic, and those where drivers are sent over a route on normal roads with other traffic present. The former is more repeatable, but less realistic. In each type of test, drivers can either be left free to drive in their normal way, or be given some training, or just be asked to drive fast, or economically.

Examples of tests on routes without other traffic were carried out by TRRL on the Laboratory's Small Road System over a route 2.6 km long, with 11 junctions and 6 stops (see Figure 4.4).

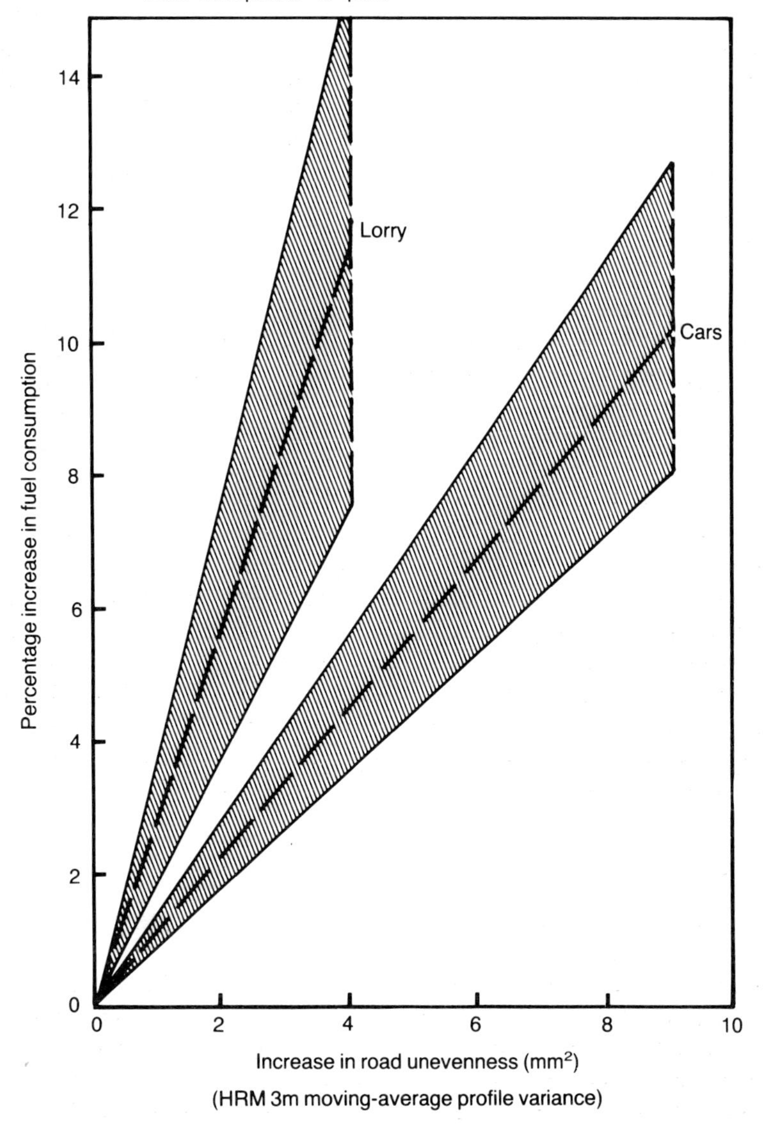

Fig 4.3 *Effect of increase in road un-evenness on fuel consumption (Source: Young, 1988)*

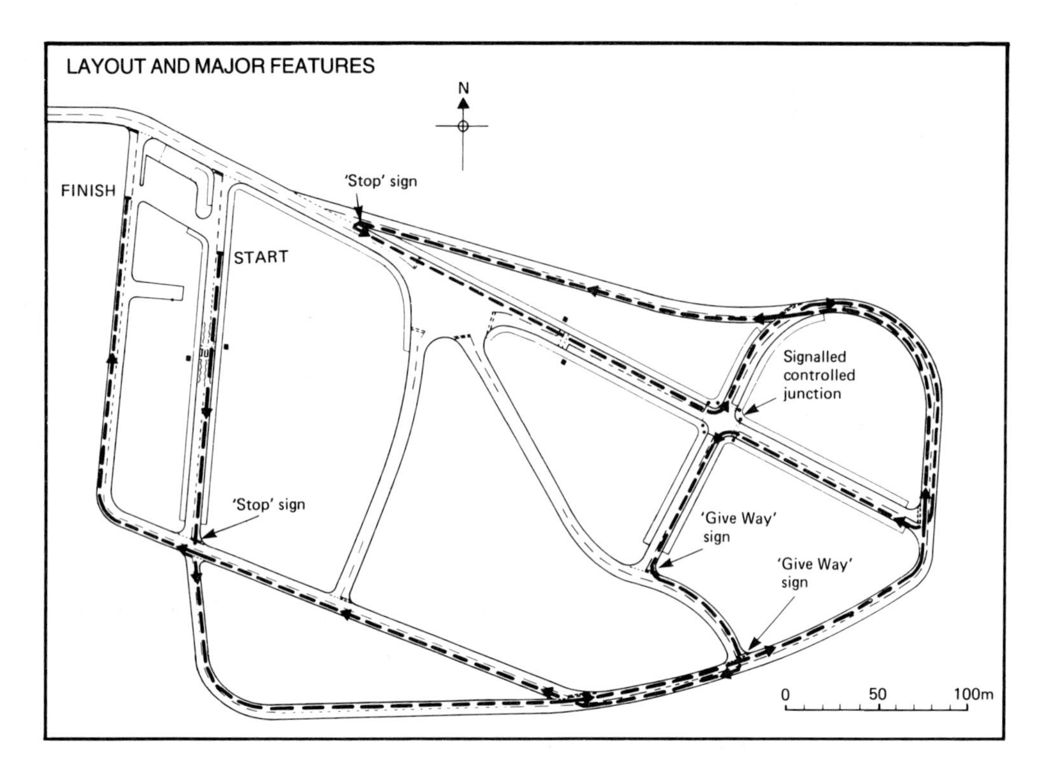

Fig 4.4 TRRL Small Road System - Test Circuit

Figure 4.5 shows the fuel consumption figures for nine different drivers, using the same car and each driving over the test route ten times. Drivers were asked to drive in a normal way. The large difference (50%) between the least and the most economical driver is notable, and so is the differing consistency of performance (from ±1.4% to ±14.0%).

A much more extensive trial was carried out over the same test route during the 1979 TRRL Open Days. Using four different petrol cars, and a diesel car, volunteer drivers were asked to drive once round the circuit normally, and once driving in a way they considered to be economical in fuel use. Some results are given in Table 4.3 showing that drivers could save more than 20% of fuel by "driving economically". Generally, the volunteers interpreted "driving economically" as driving more slowly, but when a correction was made to account for the speed differences, there was still a saving of nearly 15%, so that this saving should be possible without increased journey times.

Turning to trials of vehicles in real traffic, there is a considerable body of literature from the UK, and from other countries, and it is only possible here to pick out some of the main results. More than 20 years ago, Everall (1968) asked drivers to drive in different ways over a route in ordinary traffic. They were asked to drive "normally", "as economically as possible", and "as if you were in a hurry". For a small car, the results showed that fuel consumption and average speed were

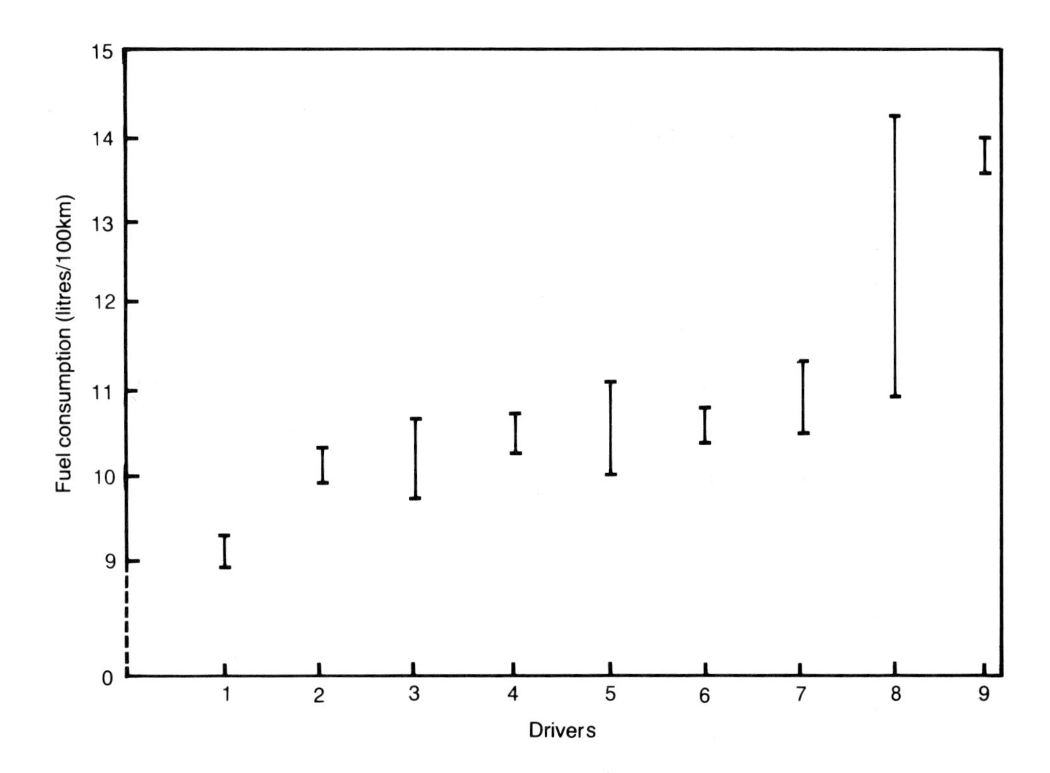

Fig 4.5 *Driver characteristics and fuel consumption*
(Source: Waters and Laker, 1980)

about 12% greater than normal when driving fast, and about 12% less when driving economically. Other trials have produced limited information about the variability of driver performance as one of the lesser objectives (see, for example: Weeks,1981; Wood *et al*, 1981).

More emphasis was placed on driver behaviour in the trials of the petrol and diesel Cavalier cars already referred to (Redsell *et al*, 1988). Six volunteer drivers and a seventh "expert" took part in the test programme. In addition to the usual measurements of fuel consumption, vehicle acceleration and deceleration and throttle position and movement were recorded. The "expert" driver was the only one who knew that the trials were concerned with fuel economy, and he received some training in economy driving. The other drivers were told to drive in their normal way.

The results obtained for the six volunteer drivers are shown in Figures 4.6 and 4.7. The result for the "expert" driver is also shown. Under urban driving conditions, fuel consumption varied by ±7% from the mean value of the six, but the corresponding average speeds were very similar (except for driver No.4 who drove faster under all traffic conditions). In suburban and motorway driving, higher average speeds significantly increased fuel consumption. Driver No.4, on the motorway, showed that with a speed 23% higher than average, his fuel consumption was 40% greater.

TABLE 4.3
Economy driving on the TRRL Small Road System
(Mean speed (km/h) and fuel consumption (litres/100 km) of each vehicle for all drivers)

Car model	Mode of driving		Mean	Standard error of the mean	Number of runs (N)		D-E % D
Allegro	Datum (D)	Speed	33.2	0.47	69	Speed	17
		Fuel	11.595	0.20	69		
	Economy (E)	Speed	27.7	0.33	69	Fuel	23
		Fuel	8.891	0.11	69		
Cortina	Datum (D)	Speed	30.8	0.40	86	Speed	15
		Fuel	13.122	0.22	86		
	Economy (E)	Speed	26.3	0.24	86	Fuel	21
		Fuel	10.397	0.11	86		
Escort	Datum (D)	Speed	31.4	0.47	72	Speed	18
		Fuel	13.807	0.22	72		
	Economy (E)	Speed	25.8	0.32	72	Fuel	24
		Fuel	10.535	0.10	72		
VW (Petrol)	Datum (D)	Speed	27.9	0.61	95	Speed	13
		Fuel	11.049	0.19	95		
	Economy (E)	Speed	24.4	0.40	96	Fuel	20
		Fuel	8.794	0.12	96		
VW (Diesel)	Datum (D)	Speed	31.0	0.37	95	Speed	15
		Fuel	8.307	0.21	95		
	Economy (E)	Speed	26.3	0.22	95	Fuel	30
		Fuel	5.84	0.07	95		

Source: Laker, 1981

The "expert" achieved a fuel consumption 9% lower than the six volunteers in urban driving, with an average speed 6% lower. Over the suburban sections, he returned a 10% fuel saving with only a 1% decrease in speed. On the motorway, he drove much slower than the others (22%) and his fuel consumption was 24% lower. Analysis of vehicle acceleration and deceleration, and of throttle position and rate of movement for the different drivers confirmed that high values of these factors correlated with higher fuel consumption.

Other analyses of driver behaviour have been carried out, notably in General Motors Research Group. Evans (1979) concluded that in urban and suburban driving, each 1% reduction (or increase) of average speed reduced (or increased) fuel consumption by about 1.1%. Evans also found that expert drivers could save some fuel without increasing trip time by adjusting their speed to avoid stops at traffic signals. Chang and Herman (1980) measured the different fuel consumption obtained by drivers instructed to drive "conservatively" or "aggressively". They found that driving conservatively reduced average speed and fuel consumption by about 10%, but driving aggressively increased fuel consumption by nearly 30% with only a 10% increase in average speed.

These, and other studies, show clearly that drivers can reduce fuel consumption without a great deal of expert knowledge of economy driving techniques. Just being asked to drive more

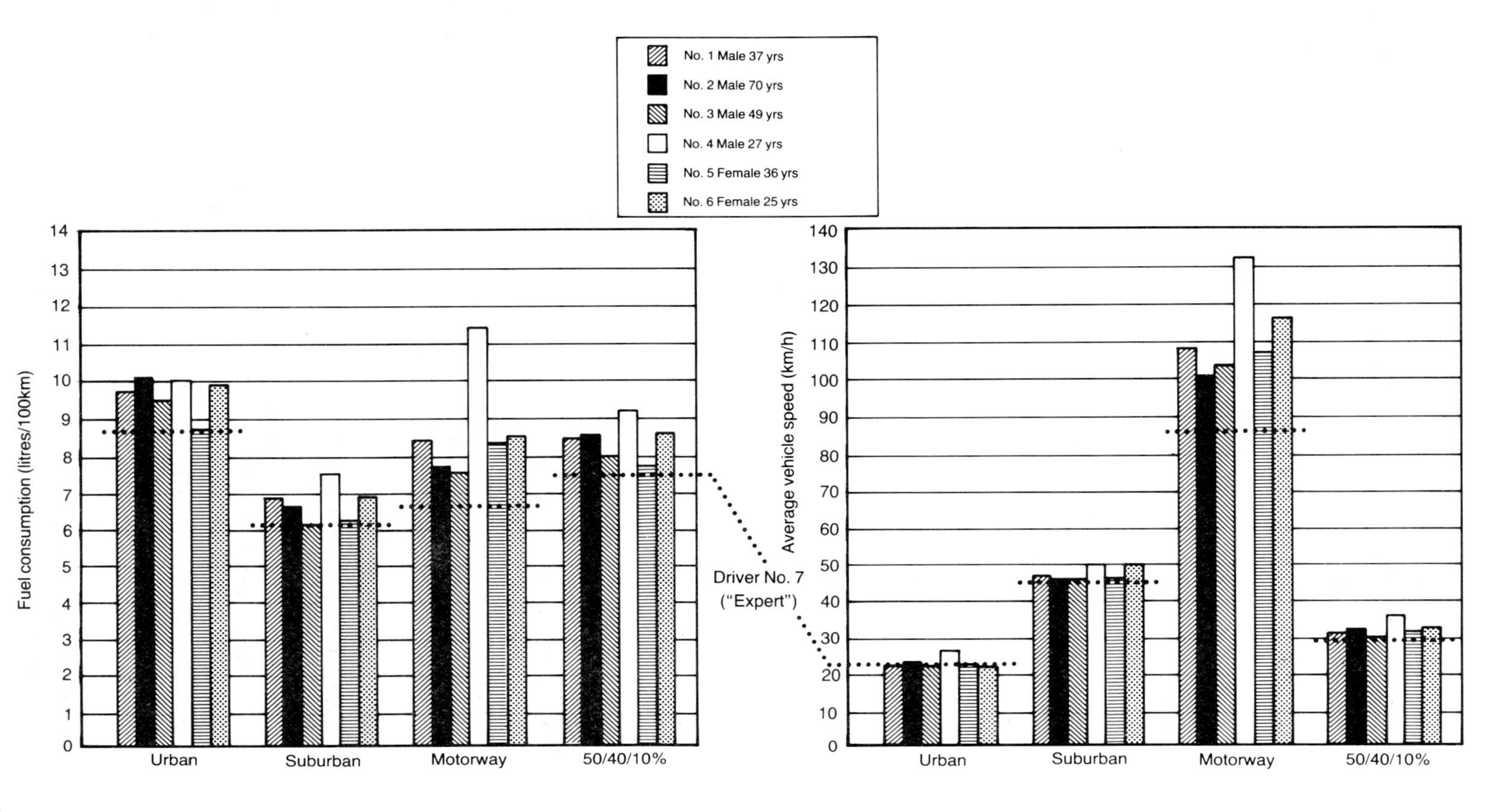

Fig 4.6/7 *Vauxhall Cavalier tests: average fuel consumption/average speed for each driver - all vehicles (Source: Redsell et al, 1988)*

economically, or more conservatively, or less aggressively produced substantial savings, though usually with some reduction of average speed. If drivers anticipated the actions of traffic signals, they could save fuel without increasing their journey time. Evans (1979) gives the following advice for reducing fuel consumption in urban traffic:-

(1) Anticipate conditions ahead so that braking is minimised. Do not accelerate to a higher speed than required if you must later slow or stop. Every time the brakes are applied, energy previously extracted from the fuel is unproductively dissipated.

(2) Avoid stopped delays. Fuel used idling while the vehicle is stopped is of great importance in urban driving. Also, after the stop, the lost kinetic energy must be restored from the fuel.

(3) Use low acceleration levels, unless a higher level will contribute to achieving (1) or (2) - for example by accelerating briskly to avoid stopping at a traffic signal. (This is not, of course, an encouragement to drive through traffic lights at red or amber.)

For car drivers on all roads, the following advice is given in the booklet New Car Fuel Consumption. The Official Figures (Department of Transport, 1990a):-

> "This booklet can help you to choose an economical car. And, once you have bought it, there is a lot you can do to conserve fuel and to help cut the cost of your motoring:

- See that your car is properly tuned and serviced according to the manufacturer's recommendations.
- Make sure your tyres are inflated to the correct pressure.
- Avoid using your car for very short journeys, especially when the engine is cold. This can easily double your fuel consumption.
- Always use the correct grade of petrol.
- Push in the choke as soon as the car will run smoothly without it.
- Watch the road ahead and try to anticipate traffic movements.
- Accelerate smoothly to a sensible cruising speed. Cruising at 70 mph can use over a third more petrol than cruising at 50 mph.
- Avoid sharp braking, which is the biggest fuel waster.
- Don't carry unnecessary weight in your car; it uses extra fuel when you accelerate or climb a hill.
- A roof rack increases wind resistance, so always take it off when you are not using it.
- Keep a check on your mpg to see if you are getting the most out of your fuel. This will also give early warning of faults that increase consumption".

In the late 70's and early 80's a number of petrol companies also produced leaflets on fuel economy for car drivers. The advice given was similar to that reproduced above.

This is good general advice on driving economically. But drivers of cars and commercial vehicles could benefit from more detailed instructions and training for achieving good fuel economy. The use of "economy meters" in the vehicle as a driver aid is a possibility. Perhaps most important, the buyer of a new or second hand car or lorry needs information about the fuel consumption of prospective purchases for the particular use that is to be made of them. The next paragraph looks at economy aids, training and information.

4.5 Economy aids, training, and information for drivers and operators

In the mid 1970's and early 80's, there was a great deal of interest in fuel economy indicators - instruments which would give the driver a continuous indication of how efficiently he was using fuel. Devices ranged from simple inlet manifold vacuum pressure gauges, to computer based indicators showing the driver the most economical gear to be in. A critical review of the devices being considered in 1980 was made in an unpublished paper (Waters, 1981).

The main drawback of the simple indicators was that some could be positively misleading. Use of the manifold vacuum gauge, for example, was reported by Evans (1979) to give only a small, not statistically significant, reduction in fuel consumption. Claffey (1979) thought that the gauge was of little use, and might cause drivers to use more fuel, not less. Ford (1982) argued that the gauge could stop drivers using the open throttle/low engine speed region which is in fact efficient for the typical petrol engine.

More complex indicators, such as instantaneous "miles-per-gallon" meters also have theoretical shortcomings, as they do not give credit for kinetic energy usefully gained. (A way of including this factor was proposed by Waters, 1981.) The general conclusion, reached in the early 80's was that the indicators were of very limited use, and drivers were unlikely to use them and save fuel. Even though the gear change indicator fitted by Volkswagen was technically sound, drivers tended to find it distracting in practice (Motor, 1981). Economy indicators are rarely fitted to production cars now, though some automatic transmissions have an "economy" and a "performance" selector to modify the gearchange logic.

In a review of energy conservation for car drivers, Naysmith (1989) came to the conclusion that a possible way forward was to change the connection between throttle pedal and engine from a mechanical linkage to a simple electronic "drive-by-wire", so that limits on throttle opening rate and acceleration could be applied. Redsell *et al* (1988) had already shown that economical drivers were characterised by low values of these parameters. Naysmith admitted that the presence of "Little Sister" under the bonnet would not be welcomed by the enthusiastic driver who likes to be able to "accelerate out of trouble" (*sic*), but in the future, if fuel prices increase substantially again, there might be more incentive for manufacturers to fit such devices. As a means of saving fuel (and also CO_2 emissions), Naysmith argued that their use could be encouraged by tax concessions.

If suitable fuel economy indicators were developed and became popular in cars, they would be one kind of training aid for drivers, because they could allow the driver to identify wasteful habits and encourage frugal ones ("learning to play by ear"). For commercial vehicles, there is more possibility of formal driver education and training, and perhaps greater potential for using driver aid devices. Booklets like those published by the Energy Efficiency Office (1987) and the Australian National Energy Conservation Program (1983) give specific advice on economical driving techniques for heavy trucks, and quote encouraging results from driver training programmes. For example, a major parcels delivery company carried out a training exercise with 31 drivers, and all returned better fuel consumption. While the variation between drivers was large, the average improvement was nearly 15%, achieved with no increase in journey time (Energy Efficiency Office, 1987). Developments of driver aids for goods vehicles include:-

- cruise speed control, and road speed governing.
- fuel consumption meters.
- gearchange speed indicators.

Even without these aids, Martin and Shock (1989) quote a study that suggests that a programme of publicity within a firm, and driver re-training could offer fuel savings of around 5%. Damongeot (1989) has reported on four case studies with both goods and passenger transport in France. He found that fuel consumption reductions from 50 litres/100km. to about 33 litres/100km. were achieved through general company policy, closely linked to training. These gains were larger than had been forecast.

One of the main sources of information for UK private car drivers about the potential fuel economy of cars is the booklet on New Car Fuel Consumption. The Official Figures (Department of Transport, 1990a), which is issued twice yearly and has already been referred to in connection with advice on economical driving. Three fuel consumption figures are quoted for each model:-

- Urban - based on a laboratory test cycle, on a "rolling road" dynamometer (see Figure 4.8).
- 56 mph (90 km/h) constant speed - based on laboratory test or on a test track.
- 75 mph (120 km/h) constant speed - for cars with a maximum design speed of at least 130 km/h.

The purpose of these figures is to give the customer information about the relative fuel economy of different models before he makes a purchase. In practice, fuel economy will vary with type of use, traffic conditions, and driving style. In addition, new cars are tested with fully warmed up engines to allow repeatability of results, so that cold starts in practical conditions will worsen fuel consumption. All of this means, as the Official Figures booklet explains, that they cannot be fully representative of real-life driving conditions, and that motorists should use the figures as a guide, rather than an absolute measure of their future fuel consumption.

Even so, the test methods, and the resulting fuel consumption figures, have been widely criticised (see, for example: Autocar, 1982; AA, 1983; The Guardian, 1982). The complaints, as might be expected, are mainly that the figures are based on test methods which do not represent driving in real traffic conditions, and that the cars are specially prepared by the manufacturers for the test, and thus not representative of the average car. There is also criticism that the tests are made with engines fully warmed up, and it has already been seen that this can lead to an overestimate of fuel economy (Chapter 3).

Similar reservations have been expressed about the corresponding US method (the EPA tests), which is based on two artificial test cycles, the City and the Highway (McNutt *et al*, 1982). The conclusion of a critical review (Hellman *et al*, 1982) was that vehicles did not achieve the EPA 55/45 miles per gallon values* because the test cycle results did not represent real traffic conditions, and the in-use vehicles were not exactly the same as those used for the EPA tests. On-road miles per gallon ranged from 5% to 25% higher than the EPA test values, with an average "shortfall" of around 15%. Poor traffic representation accounted for two thirds of the shortfall.

---

* A combination figure assuming 55% mileage is driven in conditions like the City cycle, and 45% like the Highway.

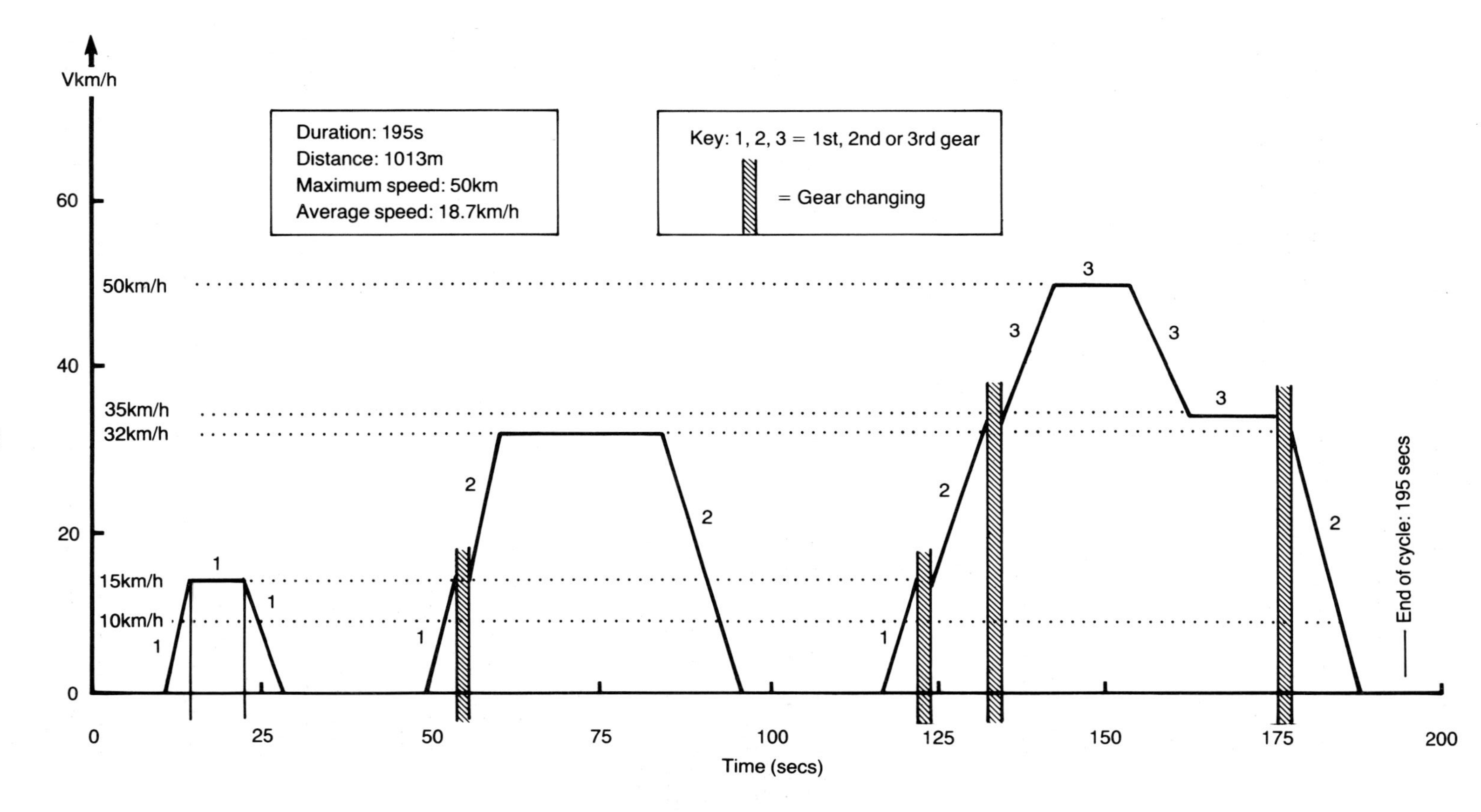

Fig 4.8 *Elementary urban test cycle (ECE 15)*
(Source: CEC, 1990)

An earlier highly critical report by a US House of Representatives Committee accused the American car industry of opposing the Government Environmental Protection Agency's efforts to make the regulatory tests more representative of real world fuel economy, and also of taking advantage of loopholes and flexibilities in the EPA test procedures to enhance the apparent fuel economy of their cars. It is little wonder that the fuel economy figures were not given much credence by the American public (US Congress, 1980).

Some of the comments about the European urban test cycle (ECE 15, Figure 4.8) may be because it has its origins as a test for exhaust emissions rather than fuel consumption. For fuel consumption testing, it has the disadvantages of gentle acceleration and braking, long periods spent stationary, and by a low average speed (less than 19 km/hr) found only in central city traffic. Gear change points are specified for a manual gearbox car: only 1st, 2nd, and 3rd gears are used. (This is why some automatic gearbox cars do well on the test cycle.) Wood (1980) concluded that, when compared with observed patterns of driving in central London, the ECE 15 cycle under-represented both low and high acceleration and deceleration and also maximum speed (Figure 4.9). Tests of different vehicles over a variety of regulatory test cycles from different countries have given not only different fuel consumption values, but have also ranked the fuel economy of the cars in a different order (Dodd *et al*, 1981). Research work on analysing actual driving in traffic and generating more realistic test cycles has been completed (Hughes, *et al* 1988), but the

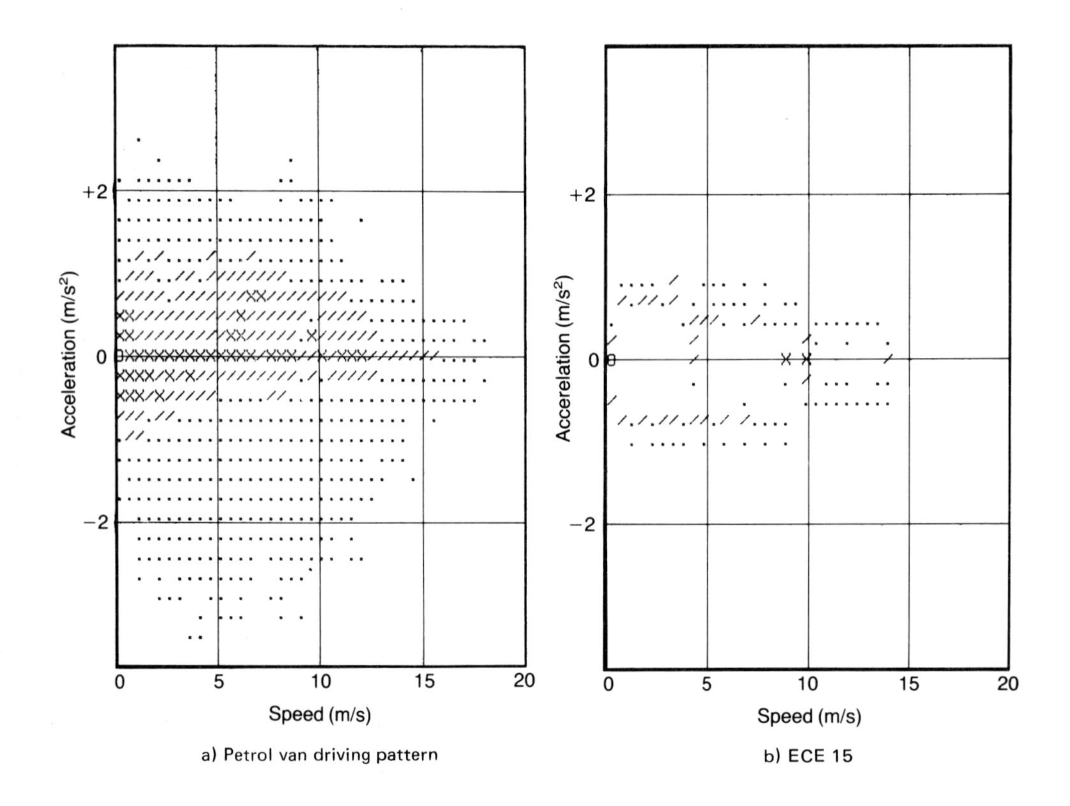

Fig 4.9 *Driving patterns of a Bedford CF petrol van in Central London traffic, and comparison with ECE 15 (Source: Wood, 1980)*

administrative difficulties of getting an improved cycle adopted by EC countries are formidable. The benefits in terms of better information are not easy to quantify, and all test cycles are a compromise in the old problem of balancing greater realism against better repeatability. Nevertheless, there has been recent progress in agreeing a new high speed driving cycle for emission testing, which could in the end be adopted for fuel consumption testing instead of the present constant speed tests (see Figure 4.10).

The most constructive comments on the present UK Official List fuel consumption figures were made by the Consumers' Association (1983), who compared the fuel consumption reported by their members with the Official List for a wide range of car models. There was a clear reduction of fuel economy (miles per gallon) with increase of engine size, as Figure 4.11 shows, both for CA's members and the Official List. When members' average results were compared with averages for the Urban, 56 mph, and 75 mph tests, it was concluded that the best guide-lines for overall fuel consumption were the Urban and the 75 mph tests. The slope of the Urban consumption in Figure 4.11 is very similar to the CA's members' experience. CA concluded that consumers should not expect to achieve figures like the 56 mph test, which were unrealistically high. It is, of course, a pity that these figures are the ones that are often given most prominence in advertisements which stress fuel economy. While the advertisers are required to give all three Official List values if any are used, the less favourable ones sometimes appear in very small print.

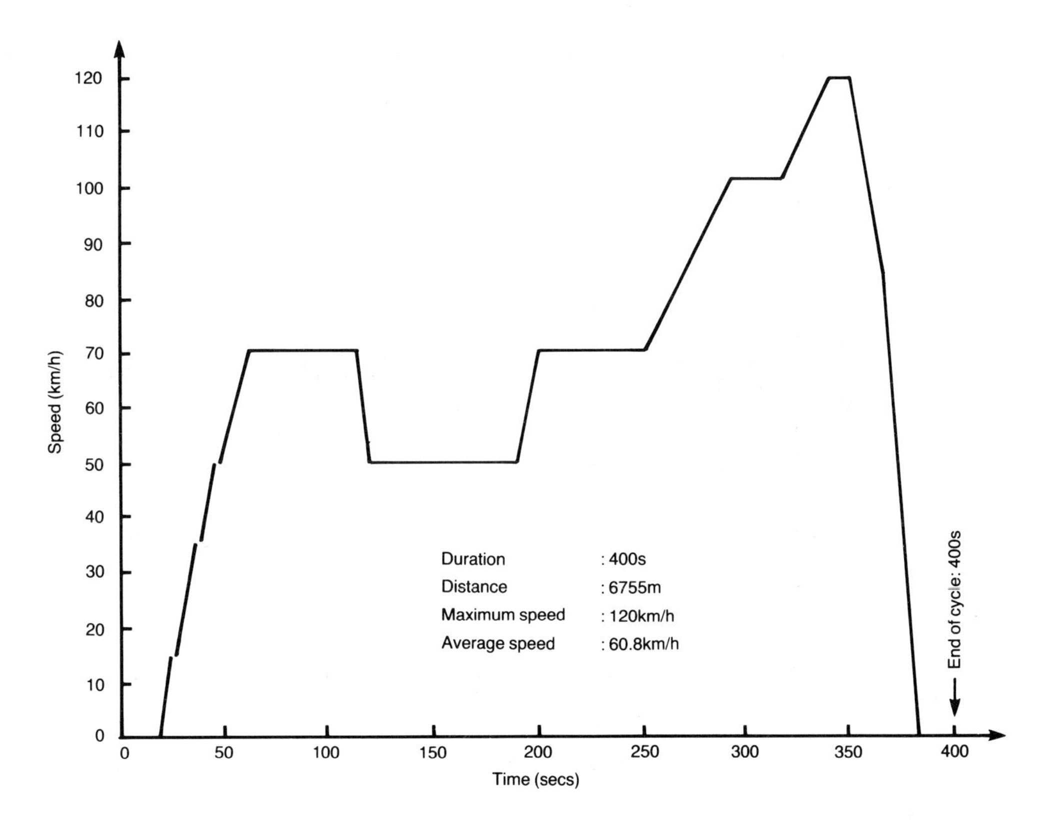

Fig 4.10 Proposed high speed urban driving cycle for emission tests (Source: McArragher et al, 1989 and CEC, 1990)

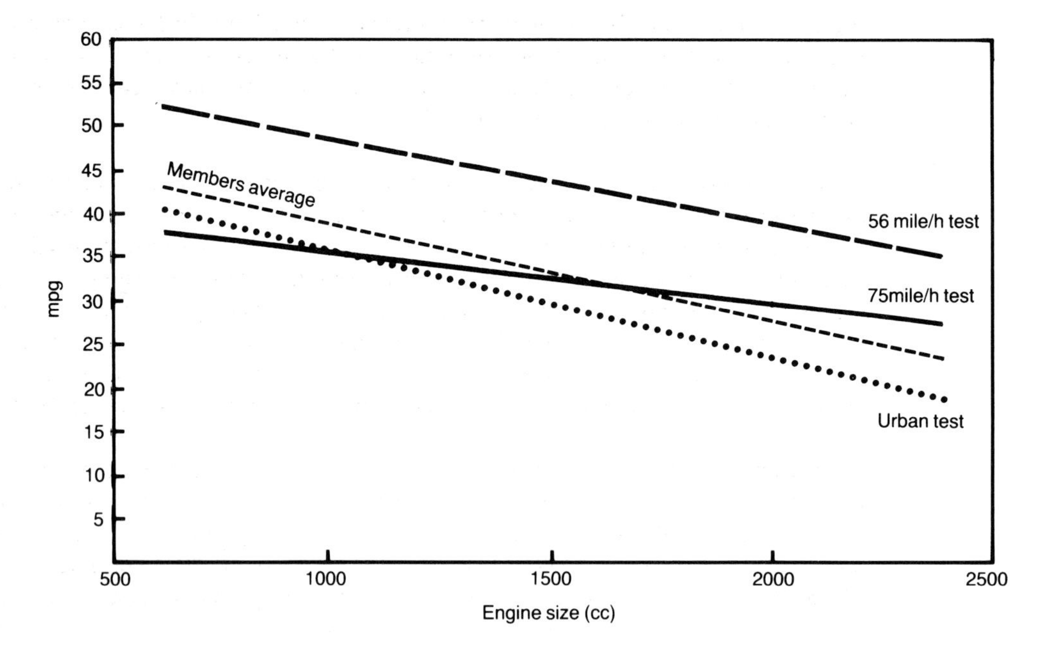

Fig 4.11 *Variation of fuel economy with engine size, and comparison with Official List values (Source: Consumers' Association, 1983)*

The emphasis on the 56 mph value, which is usually the best of the three, also tends to give the impression that 56 mph is the optimum speed for fuel economy. In fact, the motorist will get better economy driving steadily at the lowest speed possible before he has to change down from top gear.

A larger set of data from the Consumers' Association was examined for TRRL, and further analysis carried out and reported by Watson (1989). The main object was to see what combination of the three UK Official List figures could best represent overall fuel consumption as experienced on the road. A number of formulae of increasing complexity were derived, some of them including engine size as a variable. The simplest relation to give a reasonable fit to the data was found to be:-

$$\text{Fuel Consumption} = 0.60G_U + 0.26G_{90} + 0.14G_{120}$$

where: G_U = Test consumption on the Urban cycle.
G_{90} = Test consumption at a steady 90 km/h.
G_{120} = Test consumption at a steady 120 km/h.

[This may be compared with the often used combination :-

$$\text{Fuel Consumption} = 0.40G_U + 0.50G_{90} + 0.10G_{120}$$

(Department of Transport, 1984)]

A general conclusion to be drawn from the Official List and similar fuel consumption information is that it is of some value to the motorist thinking of buying a new car, but it would be unwise to expect the fuel consumption obtained in practice to be the same as any single or composite figure. At best, the Official List gives a useful guide.

There are no comparable test methods for heavy goods vehicles. Most operators who have to make decisions about purchases are competent to assess fuel consumption as one of the factors in vehicle choice, and the major manufacturers are willing and able to match the vehicle specification (eg power and gearing) to the particular route pattern of the purchaser. The technical press (eg Commercial Motor) also carry out standard tests over representative test routes, which give a better idea of comparative fuel consumption (and other matters) than the rather uncontrolled tests in traffic carried out by the motoring press on cars. (The "Scottish test route" used by Commercial Motor for heavy goods vehicles is described by Gyenes, 1978.)

A last example of information being applied to the economical use of fuel is the provision of some form of route guidance for drivers on the road network which could save fuel by reducing wasted vehicle mileage, either by not taking the shortest route, or by actually getting lost. Some early tests suggested that 4% of route kilometres were unnecessary, and that about three quarters of the wasted distance could, in principle, be recovered by improved road signs, better maps, or by providing an automatic route guidance system (Armstrong, 1977). Clearly, the extent to which the "wasted" 3% of vehicle travel could in practice be retrieved will depend on many factors, and it is not obvious that shorter routes will always save fuel (savings depend on traffic congestion, for instance). Many of these points have been examined by Jeffery (1981) who made estimates of the money savings that various degrees of guidance could produce. As a result, one proposal for electronic route guidance, "Autoguide", is under development in the U.K, and will have the potential of selecting the quickest or shortest route from information that it holds on traffic using the network (Department of Transport, 1988). In principle, it should be possible to select a lowest fuel use route, so that maximum savings of fuel are made, if this is what the driver wishes. Work is progressing in many countries, and a review of activity on route guidance and in-car communication systems was published by an international group of experts (OECD, 1988).

4.6 Gadgets to improve fuel economy

Motorists, in particular, are the target of advertising by many firms who market devices or additives which are claimed to improve fuel economy. Some of the claims of improvement are so large that the ordinary driver may well wonder whether the device should be given a try - even if credibility is strained.

Devices (or additives) tend to be grouped under a series of headings:-

- Those which weaken the mixture of petrol and air, or improve mixing before the charge reaches the cylinder.
- Those which strengthen the spark in a spark ignition engine.
- Those which are claimed to affect the fuel, either by chemical or even magnetic means.

This list is not exhaustive, but in all cases the motorist should ask whether any fuel economy is associated with undesirable effects (eg weakening the mixture may lead to misfiring and reduction in drivability), and why the vehicle manufacturer has not included the "gadget" in his own continued search for fuel economy. The introduction of increasingly stringent exhaust emission limits means that fuel and ignition controls are very carefully managed, often in conjunction with an engine management computer. It is quite likely that devices which affect the fuel/air mixture will adversely affect the improvements in emissions.

In the late 1970's many devices were offered for sale in the era of rapidly rising petrol prices. When tested by organisations such as the Consumers' Association (1978) and the AA (1981), the advice to drivers was that the gadgets were not cost effective. The Automobile Association have said more recently that the only effective device that they have tested was one which cut off the fuel from the carburettor when the engine was "over-run" (ie the engine was being used as a brake). As some vehicle manufacturers have incorporated this feature in their engine management system, the device, as an "add-on" extra, is no longer on the market (AA, 1991).

But the motorist, and other vehicle users, are still offered "goodies" which can either be bolted on or added to the fuel. In 1991, the use of tin alloys as fuel treatment received publicity as a way of allowing engines not designed for unleaded petrol to use lead-free fuel. Other advantages are claimed - as the following extracts from an advertisement show:-

> "To all motorists (from £50 plus VAT), we offer the "Fuelsave"*, together with 150,000 miles guarantee or a full refund in the unlikely event of not being satisfied.
>
> *The product*:
>
> (A) Will save you up to 30 pence per gallon (petrol or diesel)
> (B) Will reduce poisonous exhaust emissions by up to 50%
> (C)Will let you convert instantly to lead-free petrol with no adjustment or loss of power
> (D)Will give you up to 3% more power
> (E) Is fitted in minutes into fuel line of any engine
> (F) Is entirely maintenance free
> (G) Is fully operational after first 500 miles
> (H) Sales over 120,000 units
> (I) Will save you the high price of a catalytic converter
> (J) Suitable for marine craft and motorcycle
>
> Whether you are a company, fleet operator, or a private motorist, it makes sense to fit a "Fuelsave" for a smoother, quieter and more responsive engine."
>
> (Based on Motoring Exchange and Mart, 1991.)

Fuel treatments of this kind were investigated by Autocar and Motor (1991), who concluded that evidence regarding the efficiency of tin treatments was incomplete and contradictory. The reported opinions of the product manufacturers and satisfied users were contrary to the views put

---

* A fictitious product name

forward by the Automobile Association, Austin Rover, and others. The motorist is still faced with the dilemma of whether to use the treatment for the reported fuel economy benefits, but at the risk of possible engine damage.

Perhaps the best advice to give to drivers who are tempted to use fuel treatments or to buy "add-on" gadgets is still that found in Blackmore and Thomas (1977). These authors were sceptical of most of the claims made, and warned the motorist of the many ways in which devices and additives could improve economy but with corresponding losses (eg of performance) which were not always mentioned. Devices could also cost more than the fuel saved over a reasonable period: the pay-back time could be many years. Altogether, fuel economy gadgets and fuel treatments were regarded as "not proven".

4.7 Summary

The effect of other traffic on the fuel economy of a car has been represented by mathematical models of varying complexity, depending on the use to be made of them.

Provision of adequate traffic capacity to avoid congestion, and the "tuning" of traffic management schemes can both improve vehicle fuel economy in the traffic scheme. So can properly enforced speed limits on high speed roads, though savings of only a few percent are likely with realistic limits. There may be more scope for fuel savings by fitting speed limiters on the heaviest lorries which run most of their mileage on motorways.

Good standards of road maintenance can prevent fuel consumption increasing on uneven roads.

The influence of different drivers on a vehicles' fuel economy can lead to quite large changes in fuel use: 15% improvement in fuel economy can be achieved without driving more slowly. General advice on economical driving techniques has been available for many years: anticipation and smoothness of driving seem to be the key points.

Economy indicators were popular in the late 1970's, but had theoretical shortcomings, and have largely fallen out of fashion. There may be more scope for their development and use by professional lorry drivers.

The information provided by various "regulatory" fuel consumption tests is intended to be a guide to the customer, but the results are not always representative of consumption actually achieved in normal driving. There are several reasons why these test figures do not represent "on-the-road" fuel economy, and a better combination of the three EC Test Figures is suggested, to help them be accepted as a useful guide.

Drivers are advised to treat claims for large improvements in fuel economy from fitting "add-on" gadgets or fuel treatment with some caution.

5. Taxation and fuel economy

5.1 Introduction

At first sight, the connection between taxation of road vehicles, provision and design of roads, and fuel economy may seem rather remote. However, in the past, the level and method of taxation has strongly influenced the road system, has modified the design of the motor vehicle, and has affected the way the vehicle has been used. All these factors have reacted on fuel economy, though often in an indirect way: the objectives of taxation are many and varied. Some of these fiscal effects are still present today, and a brief look at the history of vehicle taxation is useful, both because it provides a sense of perspective, and because it contains lessons which are still relevant. This Chapter therefore starts with a review of the earliest methods of raising revenue from road motor vehicles, and goes on in later Sections to examine the effects of taxation on vehicle purchase, ownership, and use.

The early use of the motor car in the UK at the turn of the century gave rise to three main areas of public concern: danger; dust; and cost*. The danger element is obvious; fatal and less serious accidents were occurring with increasing frequency, and an attitude of antagonism was developing between the motoring fraternity (seen as rich, arrogant and inconsiderate) and non-motorists.

The problem of dust has now been almost forgotten, but in the early 1900's it caused most of the complaints against the motor car, especially from people in rural areas where crops and gardens had been ruined by dust thrown up by motor cars from dry, unsurfaced roads. Many patents for "dust-free" cars were filed, but these were not successful, and the solution was obvious, though expensive. It was to tarmac all the road surfaces. As, by 1909, there were only 53,000 motor cars and light vans in use (out of a total of 144,000 motor vehicles) and as the length of roads was about 280,000 km (Department of Transport, 1990a), it was not at all evident that this was a cost effective solution. More seriously, it raised all the old problems of who should pay, which had bedevilled the funding of road construction and maintenance back to the age of turnpike roads and beyond. Local authorities objected to spending ratepayers money on roads which were used by "outsiders" on long through journeys. There was no national roads design and construction organisation of the kind that is now taken for granted, and considerable resistance to the idea of one being set up.

After Byzantine negotiations with the pro- and anti-motor car lobbies, the Chancellor of the Exchequer (Lloyd George) announced, in his Budget Speech in 1909, the setting up of a new central road board to oversee road improvement and new construction, a tax on vehicle ownership (based on the RAC horse-power formula, as a proxy for vehicle weight) and a tax on

* For the historical background, the author is indebted to Buchanan (1958) and Plowden (1971).

petrol used in motor vehicles - at the rate of 3d per gallon for imported* petrol (and with a 50% rebate for commercial vehicles). All the extra revenue from the new taxes was to go to the Road Board and be used for improving, constructing, and (after a House of Lords amendment) maintaining roads for motor traffic.

This was, for the UK, an extraordinary proposal to reserve the proceeds of one form of taxation for one particular purpose. The pledge given to motorists by Lloyd George in 1909 was, again extraordinarily, repeated by the then Chancellor (Austen Chamberlain) after the Great War in 1919. As Plowden (1971, page 94) comments, the eventual failure of the Road Fund was not so striking - "given the heretical taxation principles on which it was based" - as its survival into the mid-1930's in practice, into the mid-1950's in form (the Road Fund Licence for vehicles), and into the 1960's as a weapon of political argument for increased spending on roads. It was, incidentally, not until Churchill, in a robust mood as Chancellor of the Exchequer, drove through the first "raid" on the finances of the Road Fund in 1926, that this most unusual assignment of specific taxes to specific expenditure began to be eroded - greatly to the relief of the Treasury, who had resisted the principle of hypothecation from the start.

This historical account has several relevant pointers as far as present day fuel economy is concerned. First, the formation of a central organisation, responsible for constructing and improving and maintaining trunk roads, had its origin in the 1909 Road Board. The UK Department of Transport now has this function amongst its responsibilities, but no longer on the basis of all funds raised from motor vehicle taxation. However, the design standards of roads, and their safety and capacity are centrally organised countrywide, and design guidance given to local authorities for those roads which remain their responsibility, or are managed by the authorities as agents for the Department. Any changes in design to improve road traffic fuel economy can be developed and applied on a national basis.

Secondly, the principles of motor vehicle taxation gradually evolved from the 1909 Lloyd George proposals of an ownership tax (based on rated horse-power and a use tax (based on fuel used), through the mid 1920's principles suggested by Churchill of taxation by wear and tear of the road plus an element for "luxurious locomotion" (Plowden, 1971, page 200), towards today's (UK) system of a percentage tax on the purchase of new cars (Car Tax) a flat rate ownership tax for cars and light vans (Vehicle Excise Duty), a high element of tax in the price of petrol or diesel fuel, and an annual tax on heavy goods vehicles based on estimates of road structural damage due to the varied axle loads of different classes of HGV (Department of Transport, 1990b).

The period of the horse-power tax for cars, which lasted up to 1947, introduced engine design limitations which influenced fuel consumption, but also affected the export market for British designed cars. The influence on engine design is discussed in more detail later. For both cars and commercial vehicles, the incidence of fuel tax has had a considerable effect on vehicle design and use, as international comparisons demonstrate. This topic, too, is discussed more fully later, and is particularly important in relation to fuel economy, both because of the elasticity of vehicle use to fuel price (See Chapter 2), and also because of the pressure exerted towards the design and marketing of economical vehicles by the need to make good use of expensive fuel.

---

* The restriction to imported petrol was partly to placate motorists, and partly to encourage home production from inland oil fields.

Finally, there is the effect of subsidy (by reduced taxation) for the business use of vehicles. In 1909, half the fuel duty was rebated to commercial users. After the 1990 Budget, commercial (and private) users of diesel fuel paid rather less duty than users of petrol (86 pence per gallon, compared with 102 pence). There was also a full rebate of fuel duty (on both derv and petrol) to operators of eligible local bus services (Department of Transport, 1990a). Value Added Tax is, of course, repaid to business users. Moreover, the tax treatment of private cars used for business purposes - the so-called "company car" - also affects the market for cars and therefore their overall fuel consumption during their lifetime. This is discussed, as far as it relates to fuel economy, later in this Chapter.

Naturally, the intentions and effects of taxation have a much wider scope than simply fuel use. This should be borne in mind in the discussions in this Chapter which, because it concentrates on fuel efficiency, is only attempting to illuminate one facet of a very complex topic.

5.2 Taxes on vehicles

As outlined in the introduction to this Chapter, the main taxes on vehicles in the UK can be classified under three headings:

- Taxes on purchase of vehicles (eg. the Special Car Tax, currently charged at 10% of wholesale price).
- Taxes on ownership (eg. the flat rate Vehicle Excise Duty for cars and light vans, or the graduated annual tax for heavy commercial vehicles).
- Taxes on use (eg. duty on petrol or diesel fuel used for road vehicles).

In addition, Value Added Tax is charged on purchases of vehicles and fuel, but can be claimed back for business use.

The different taxes would be expected to affect vehicle fuel economy, through vehicle design and owners' choice, in differing ways. They are not, of course, levied with this specifically in mind: they are part of general taxation for revenue purposes. But in a climate where improved fuel economy is regarded as important, the effect of the present tax regime on fuel use is relevant, and possible changes in the balance of different taxes could be considered, even without changing the total revenue yield. The UK Government has stated that, to make people more aware of the environmental impact of their transport decisions, it will consider whether further changes should be made in the taxation of fuel and vehicles which might encourage people to seek greater fuel economy ("This Common Inheritance", 1990).

The revenue from different forms of vehicle taxation, and the changes between 1979 and 1989, are shown in Table 5.1. Fuel excise duty is the largest source of revenue, increasing from 44% of all road-related taxes in 1979 to 47% in 1989. Value Added Tax is the next largest item, though it is, of course, only a direct tax on the private use of motor vehicles because business and commercial users can claim a refund. (For this reason, as will be seen later, it is sometimes not included in tax on road vehicles, but regarded as part of general taxation.) Revenue from vehicle licences represented in 1989 about 16% of the total, having fallen about 4 percentage points from the 1979 proportion.

TABLE 5.1
UK road taxes - 1979 and 1989

	1979		1989	
	£M	*(%)*	*£M*	*(%)*
Fuel excise duty	2487	43.7	8457	47.0
Vehicle licences	1118	19.6	2873	16.0
Driving licences	12	0.2	22	0.1
Car tax	515	9.0	1544	8.6
Value Added Tax*	1495	26.3	4907	27.2
Import Duties	65	1.1	200	1.1
All taxes	5692	100	18003	100
of which:				
Cars, taxis and hire cars	4560	80.1	14875	82.6
Road freight	1035	18.2	2896	16.1
Buses and Coaches	97	1.7	232	1.3

* Includes only non-business VAT which is not rebated
Source: Department of Transport, 1990a

Table 5.1 also shows that most of the revenue comes from cars, taxis, and hire cars (83%), though it has to be realised that including VAT increases the tax revenue from private as opposed to commercial use.

Another breakdown of tax taken from road transport is made in the calculation of the contribution of each class of vehicle to the costs they impose on the road system. Details are published annually by the Department of Transport (1990b) and Table 5.2 gives the estimated road taxation revenue and road costs* for 1990/91. The division into vehicle class is different from that in the lower part of Table 5.1, because it has to be consistent with classification for traffic estimation on which attributed road costs are based. Table 5.2 also excludes VAT, and car tax, on the grounds that they are both part of general taxation (though this encourages the old "Road Fund" belief that there is something special about other vehicle taxation).

The proportion of tax (as defined for Table 5.2) obtained from cars, light vans and taxis is about 80%, compared with about 16% from goods vehicles.

Fuel excise duty (from all vehicles) now accounts for 75% of tax, with the remaining 25% coming from Vehicle Excise Duty. The proportions are not very different for the two main vehicles classes: for cars, fuel duty is 75%; for HGVs, 70%.

The last point to make, from Table 5.2, is the road costs attributed to the vehicle classes. Cars, etc, impose twice the cost of HGVs - a consequence of their great number and their effect on policing and traffic management costs, rather than road damage. On the other hand, cars, etc, contributed more than three times their costs in taxation, compare with only about 30% more for HGVs. These basic present day figures are useful in considering the effects of taxation on the encouragement of fuel economical vehicles.

* The road costs include the costs of maintenance, policing and lighting, but not the wider external costs of air pollution, noise, congestion and accidents.

TABLE 5.2
Estimated road taxation revenue and road costs in 1990/91: by vehicle class

Vehicle class	Number of vehicles thousand	Road taxation revenue and road costs (£ million at 1990/91 prices[1])					Taxes to costs ratio
		Road taxes			Road costs	Taxes less costs	
		Fuel tax	VED	Total			
Cars, light vans and taxis[2]	22622	7330	2385	9715[3]	3075	6645	3.2:1
Motorcycles	939	35	15	55	20	35	2.7:1
Buses and coaches	76	235	25	260[4]	220	40	1.2:1
Goods vehicles over 3.5 tonnes gvw	471	1375	580	1955	1510	445	1.3:1
Other vehicles[5]	1281	170	10	180	75	105	2.4:1
All vehicles	25389	9145	3020	12170	4895[6]	7275	2.5:1

1. Rounded to the nearest five
2. Include goods vehicles under 3.5 tonnes
3. Excludes car tax, expected to raise £1.5 billion in 1990/91
4. Includes fuel tax rebate
5. Crown, disabled and other vehicles exempt from VED, haulage machines, 3-wheeled motor vehicles, special types, recovery vehicles and non-plateable vehicles
6. Excludes £387 million allocated to pedestrians
(Source: Department of Transport, 1990b)

5.3 Tax on vehicle purchase

An example from tax on the purchase of vehicles can serve as an illustration. In New Zealand, after the 1974 oil price rise, a steeply graduated sales tax was introduced, rising from 30% on cars with engines up to 1350cc to 60% of cars with engines larger than 2700cc. The effect on the median engine capacity of new cars was much greater than the simple effect of higher petrol prices, as the international comparison in Figure 5.1 shows. Between 1975 and 1981, the average engine capacity for new cars dropped nearly 20% from 2000cc to 1680cc. The New Zealand Ministry of Energy believes that the graduated sales tax was a significant factor (IEA, 1984).

However, while tax on purchase had, in this example, a significant effect on engine size, the influence on fuel economy is indirect because fuel economy can vary markedly for the same size of engine: a direct link with fuel economy might be preferable, if that is the desired objective. And there may also be disadvantages in a high sales tax, because it may inhibit the purchase of new, more efficient, car designs so that improved technology for both economy and emission control is slow in being taken up. The IEA Report comments that, in three countries with high sales taxes, the car fleets were comparatively old.

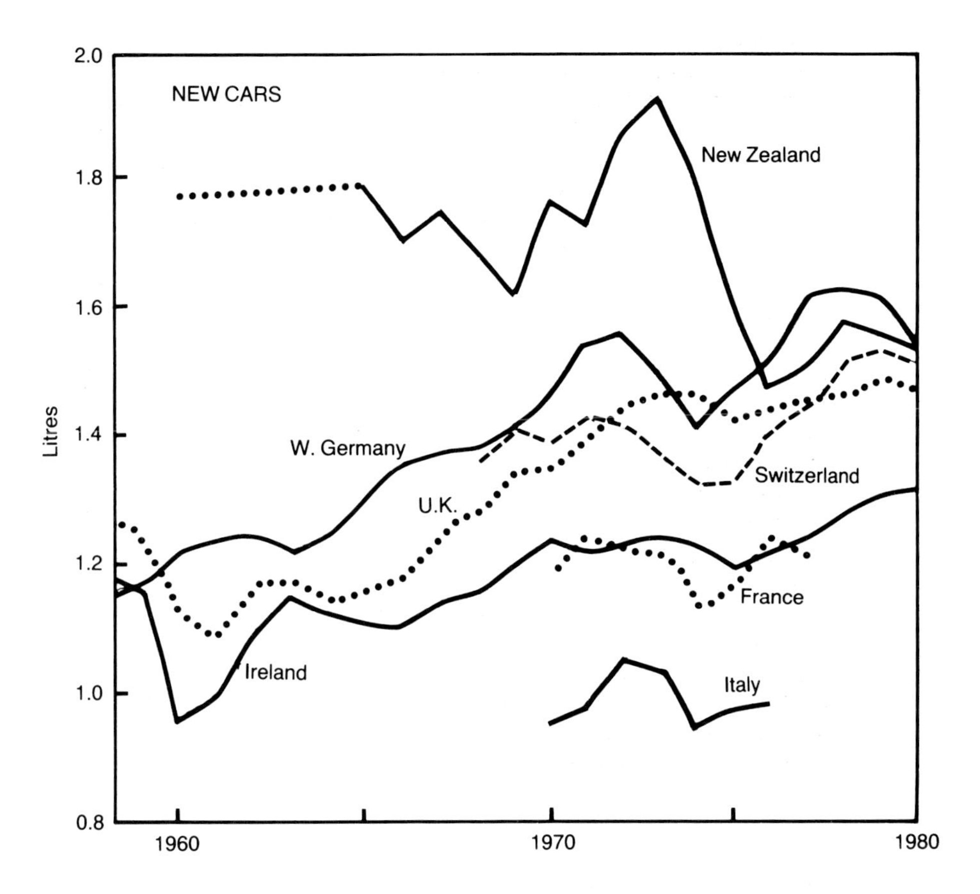

Fig 5.1 *Median engine capacity of cars in selected countries, 1958-1980 (Source: Tanner, 1983)*

5.4 Tax on vehicle ownership

An annual tax on car ownership has been a feature of the UK fiscal framework since 1909. The tax was originally based on the nominal maximum horse-power of the car engine, as a proxy for weight, which was (in those days) thought to be a measure of damage to unsurfaced roads and the generation of dust (then a serious nuisance). The formula used in calculating nominal horse-power had been developed by the Automobile Club (later the Royal Automobile Club) in 1906 as a guide to buyers of cars when comparing prices (Plowden 1971, page 167).

The formula is quoted (Newcomb and Spurr, 1989) as:

$$\text{RAC Rated HP} = 0.4\, d.^2.N$$

where: d = cylinder diameter, in inches
N = number of cylinders

The formula assumes a standard brake mean effective pressure (bmep) and piston speed for all engines. In the non-metric units of the time, these were 67 pounds per sq in and 1000 ft per min respectively. Note also that the rated HP is proportional to total cylinder area, not capacity (or swept volume), and that to increase the actual engine power for the same (taxed) rating, the designer would favour a high bmep and high speed (rpm, and thus piston speed) while using small diameter cylinders with long stroke (to get the required swept volume for power output).

The horse-power tax, as used in the UK until 1947, was therefore a design-based regulator, rather than a performance-based one, as it would have been, for example, if the tax had been linked directly to the maximum brake horse- power produced by the engine. The horse-power tax thus influenced the design of engines, sometimes in unexpected ways.

In the 1920's and 1930's, a trend towards small, light, and small engined cars was encouraged for the mass market in the UK, in contrast to the much larger engined cars being produced, for example, in the USA. The British cars were, almost incidentally, very fuel economical compared with their American counterparts, but the economy came from small, light bodies and small engines (giving modest road performance) and not from thermodynamically efficient engine design.

An example of the small mass-market car, designed for the era of the horse-power tax, was the 1928 Austin 7. It weighed just under 1000 lb (440 kg) and, on road test, returned a fuel consumption of 42.5 miles per gallon, but with a top speed of only 47 mph. Its 750 cc engine produced about 10 brake horse power (Daniels, 1970). By comparison, its RAC rated HP was 7.7, and the engine stroke was nearly 1.4 times the cylinder diameter.

Modern small petrol engines are very different, though it is not easy to disentangle the different elements of engine design which have improved engine efficiency over the years. Taking as an example the Rover 'K' series engine in standard form, it has a capacity of 1100cc and produces 60 brake horse- power for a RAC rated HP of 13.9 (Autocar and Motor, 1989). The bmep is nearly 120 pounds per sq in and the piston speed about 2,500 ft per sq in (compared with the values of 67 and 1000 used in the RAC formula). Its cylinder diameter is greater than its stroke, which gives a more thermally effective cylinder design - though it would, in the old days, have paid a penalty in nominal horse-power tax. Much of the increased power from a given capacity (B.H.P. per litre is 55 for the 'K' engine, compared with 13 for the Austin 7) comes from increased compression ratio, better valve design and the availability of higher octane fuel. But some improvement also comes from the relaxation of the (taxation) need for small diameter, long stroke and high revving engines.

Plowden (1971) concluded that the horsepower tax led, in the UK, to a rather special design of low-powered car, which had difficulty in selling in overseas markets. On the other hand, the high tax on large engined cars effectively protected the UK industry from cheap American imports like the Model T Ford, and from cheap mass production in the new works at Dagenham of the Model A Ford (which was not a success). Both these models had engines with large capacity and large cylinder diameters. They were powerful, slow revving and with good pulling power (torque) and low engine speed - characteristics which made them easy to drive. But they paid too high a rated horse-power tax to make them attractive in the UK.

A lesson to be drawn from the operation of the UK horse-power tax is, perhaps, to beware of the unpredicted effects of an attempt to tax heavy cars by an indirect tax system. It is at least arguable that there would have been less distortion of engine design, with complex consequences, if weight had been taxed, instead of rated horse-power.

Other countries have other ways of taxing car ownership. Table 5.3 shows the situation in several countries as it existed in 1984. The UK had then (and still has) a flat rate annual tax on cars and light vans, which was introduced in 1948, after a short period when tax was based on engine capacity. In other countries a wide variety of methods of assessing tax on car ownership are evident, and sometimes a combination of factors is used.

For example, Belgium and Italy used both cylinder capacity and a nominal (rated) horse-power. Australia and Japan used cylinder capacity and vehicle weight. Again, it is arguable that if the objective is to improve fuel economy, a direct measure of fuel consumption could be included in the method of assessing tax. The regulatory tests for fuel economy (eg for urban, rural, and motorway driving) which have already been described could form a basis for an 'economy' tax, but there might well need to be harmonisation of methods across Europe, especially with the approach of the Single Common Market in 1992.

TABLE 5.3

Basis of annual car ownership taxes

Country	Cylinder Capacity	Fiscal Horsepower	Weight	Fixed Rate	Other
Australia	x	-	x	-	-
Austria	x	-	-	-	-
Belgium	x	x	-	-	-
Canada	-	-	-	-	x(1)
Denmark	-	-	x	-	-
Germany	x	-	-	-	-
Greece	x	x	-	-	-
Ireland	x	x	-	-	-
Italy	x	x	-	-	-
Japan	x	-	x	-	-
Luxembourg	x	-	-	-	-
Netherlands	-	-	x	-	
New Zealand	-	-	-	x	
Norway	-	-	-	x	-
Portugal	x	-	-	-	
Spain	x	x	-	-	-
Sweden	-	-	x	-	-
Switzerland	x	-	x(2)	-	-
Turkey (3)	-	-	x	-	x
United Kingdom	-	-	-	x	-
United States	-	-	-	-	x(1)

(1) Basis for registration fee varies by state or province.

(2) Four Cantons only

(3) Since 1978 import of vehicles above 1700cc is prohibited. The motor vehicle purchase tax is based on weight and age of the car.

Source: IEA, 1984

A final point, for cars, is the level of ownership tax which might have to be levied in order to influence the purchaser's choice of model. At present, in the UK, the ownership tax is an annual flat rate Vehicle Exercise Duty (VED) of £100. For the average private car user, his annual petrol bill may be around £500, and depreciation each year on the capital value of his vehicle could be of the order of £1000. VED is not therefore the major factor in motoring costs. It could be that very substantial increases in VED for some uneconomical models would be needed before there was a significant effect on customer choice.

For heavy goods vehicles, the annual ownership tax is the vehicle licence, which is not levied at a flat rate in the UK. It is based on a formula which is intended to ensure that each class of vehicle (including cars and light vans) contributes from its tax yield at least the cost that it imposes on new construction of roads, and maintenance of existing roads (see Table 5.2). The main structural damage to roads is held to be caused by the load carried by each axle, raised to the fourth power. As HGVs have very much higher axle loads than cars, they are expected to make the major contribution to the repair of the damage, and the annual licence fee reflects this. For example, an articulated HGV with a gross weight between 36 and 38 tonnes, and with 5 axles, would pay (in 1990) an annual licence fee of £3100. If it had 6 axles, with lower loading on the road, the fee would be reduced to £1240 (Department of Transport, 1990b).

The annual tax on HGV's does not include any factors (like engine capacity) which could be regarded as a proxy for fuel use. The influence of taxation on fuel economy is therefore directly through the duty and tax on fuel which is the subject of the next Section.

5.5 Tax on vehicle use

The main source of tax revenue from vehicle use in the UK, and the one of the most interest here, is the tax (or excise duty) on fuel used for road vehicles. In other countries, toll motorways and bridges may be a significant source of revenue from vehicle use, and it could be considered that parking charges are a form of tax on use. There has also been great interest in methods of charging for the use of congested roads. Research and development work on electronic means of road pricing was carried out at TRRL in the 1970's (Charlesworth, 1987), and proposals for the use of a similar system in Hong Kong were made, but not carried beyond a pilot assessment. Another pricing system, based on delays encountered at peak times, is being considered for application in the city of Cambridge. Perhaps the most successful scheme is the one introduced in Singapore city in 1975 which requires drivers to buy special area licences before entering the central area (Hamer, 1991). While the principle of charging for road use has not found favour in the UK in the past, the Secretary of State for Transport has recently announced his intention of commissioning wide-ranging research into urban traffic congestion - its causes and possible cures. He has decided that the research should include an assessment of the possible role of road pricing (Department of Transport, 1991). However, where the interest is in encouraging fuel economy, it is reasonable to concentrate on the most direct tax - on road vehicle fuel.

An illustration of how attitudes change over time is given by the history of the fuel tax for the motor car. Introduced in 1909, and increased during the Great War, it was abolished in 1920 in favour of a system of vehicle ownership taxes. When Churchill, as Chancellor of the Exchequer, was considering and preparing his "raid" on the Road Fund in the mid 1920's, he was attracted to the reintroduction of the petrol tax as a rational tax on vehicle use. There were, however, at that time considerable technical objections, mainly concerned with the difficulty foreseen of taxing fuel for vehicles, without taxing fuel like paraffin used then for home heating and lighting.

From the perspective of the 1990's, these objections seem strange, as road fuel tax is now seen as a fair and cheaply collected source of revenue, with only limited possibilities for evasion. The addition of a dye to non-road use kerosene has made detection of misuse easier. In the end, in 1929, petrol tax was re-introduced (at 4d per gallon) in addition to vehicle duty (Plowden, 1971), and has become a major element of the taxation scene since then.

In Great Britain, taxation of fuel has been high in recent years. Over the past 15 years, the proportion of duty and tax in the retail price of petrol has been between 44% and 68%. Over the same period, the proportion in diesel fuel has been between 44% and 63% (Department of Transport 1990a).

International comparisons are difficult, because of changes in exchange rate, but Figure 5.2 shows the retail price of 4 star petrol in Italy compared with the UK from 1975 to 1990. The tax rate on petrol in Italy is even higher than in the UK, and this is reflected in the higher retail price.

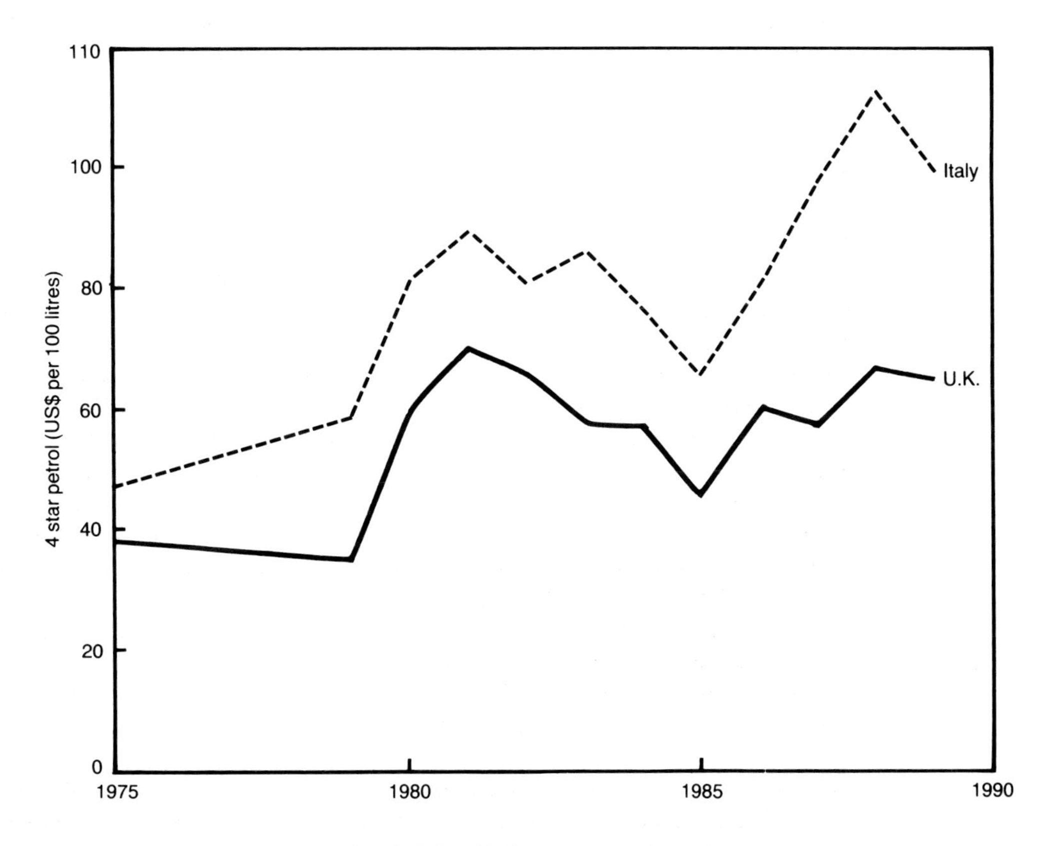

Fig 5.2 *Retail price of 4 star petrol in the UK and Italy (Source: Department of Transport, 1990a)*

Fig 5.3 *Retail price of diesel road fuel in the UK and Italy*
(Source: Department of Transport, 1990a)

By contrast, the tax on diesel fuel in Italy was low up to the mid-80's, and the effect on retail price is shown in Figure 5.3. It appears that changes in tax policy in Italy since about 1985 have been made deliberately to increase the price of diesel fuel. In 1982, the proportion of tax and duty in Italian diesel fuel was only 17%: by 1989 it was 63% (Department of Transport, 1990a). From the discussion on elasticity of demand to price in Chapter 2, it would be expected that different levels of fuel price would influence, not only the amount of travel, but also the type of car - high fuel price encouraging the purchase of economical cars. This is shown clearly in Figure 5.4 for the period 1966-1970 where the median engine size of new cars is plotted against retail petrol price. The small engine size of cars in Italy contrasts with the very large average engine size in the USA, and to a lesser extent in New Zealand (before the imposition of the sales tax).

For a thorough analysis of car ownership and use and comparison between different countries, Tanner (1983) is a valuable source. It has only been possible to summarise a few of his conclusions here. It is worth reporting one of them in full:

> "The patterns that emerged...are certainly only crude approximations to the complex interactions of cause and effect that operate in such a diverse set of countries. The fact that they did emerge at all perhaps illustrates that despite historical and cultural differences, people to a great extent respond to economic forces as expressed in prices and incomes. This does not mean that there is anything inevitable about the extent to which people run or use cars; on the contrary, the evidence shows how different policies, especially in respect of taxation and prices, can lead to different results."

This conclusion suggests that, if an objective is to encourage economy in the use of road transport fuel, taxation offers one means of achieving this. The research evidence suggests that tilting the

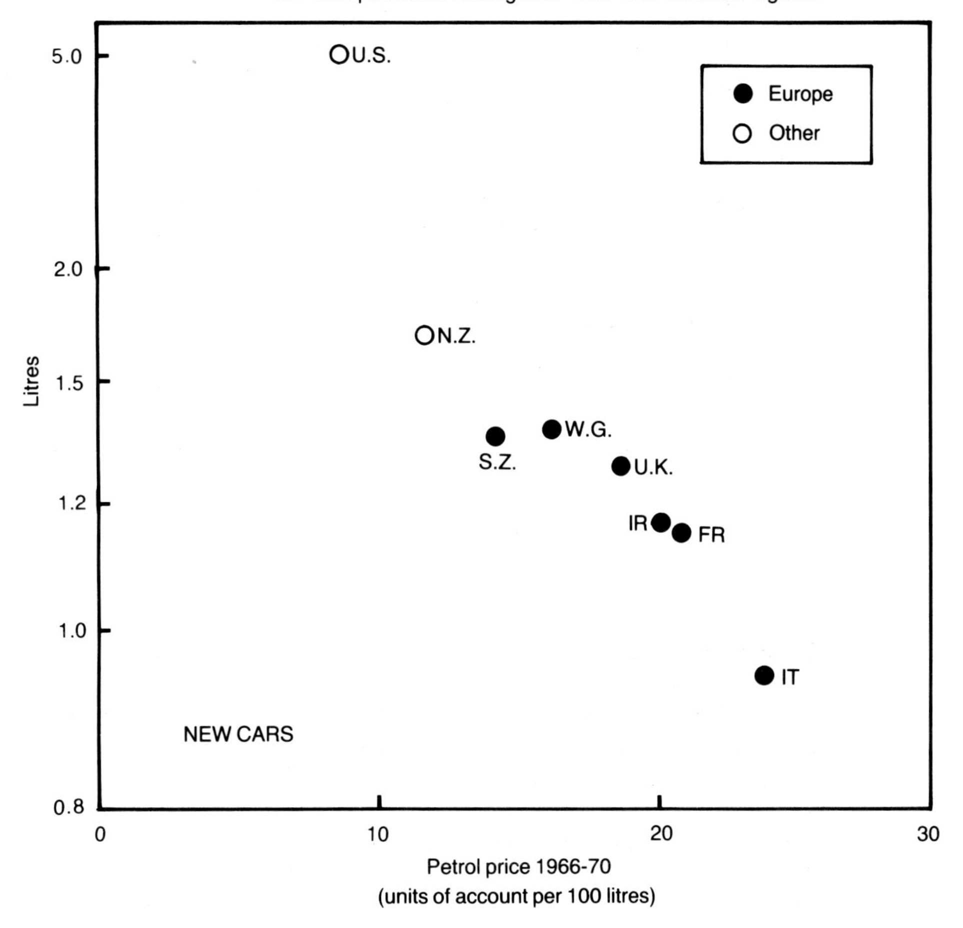

Fig 5.4 *Median engine capacity of new cars, 1966-1970, and petrol price (Source: Tanner, 1983)*

balance of taxation even more towards fuel tax, and away from purchase and ownership taxation, would further encourage fuel economy in road vehicles.

5.6 Company cars

Naturally, policies affecting business taxation are primarily concerned with the wider commercial performance of companies, and the issue of company-owned or subsidised cars is a small item in these larger considerations. Nevertheless, financial support from companies for business motoring has appeared to influence the market for cars in the UK substantially, and it is therefore a factor to be considered when trying to understand why the UK car fleet is as it is, and what might be done to encourage greater fuel economy.

A thorough study of the transport implications of company-financed motoring was carried out by Hopkin (1986). Unfortunately, the data relate to a time when the level of tax on company cars was much lower than it is now. Thus, while this provided a good opportunity to study the effects of motoring costs on behaviour with regard to car purchase and use, the conclusions do not reflect the current situation, nor does the study consider other fiscal measures that could influence this behaviour, though some consideration has been given in previous Sections to effects of taxation on fuel economy.

> "Although some company cars are tools of the trade, many are 'perks' allocated to employees with higher status and income. One survey of executives showed that in 1982, 99% of managing directors and 95% of sales executives had a company car, but in some specialist jobs the proportion was only 34%." (Hopkin, 1986).

The combined effect of low tax on company car benefits and wage constraints in the early 1970's made the company car increasingly popular for employees, and consequently the proportion of new cars registered in a company's name has risen from 37% in 1975 to 51% in 1988 (Open University, 1989). In 1978/79, it was estimated that 31% of privately-taxed vehicles available for private use received some form of financial support from companies, although only one third of them were wholly or partly company owned (Hopkin, 1986). Company support for private cars, in various forms, is clearly widespread in the UK.

The main limitation of Hopkin's study is that it is based on data from the National Travel Survey 1978/79, and the Survey was not designed for this purpose. For example, it is not clear whether hire and pool cars are fully represented. Furthermore, the levels of tax on company car benefits were less than a third of 1990 rates. More recent survey work was carried out on company cars in the London area by TRRL in 1989 but the results are not yet published. The Hopkin study is thus the most comprehensive analysis currently available and, while the data may need updating, it does deal in depth with the possible consequences for transport, including such factors as:

(a) People who are given a company car instead of a pay rise may keep their own car as well, and this may tend to increase overall car ownership.

(b) The company car holder may travel further and more often than if he met the full cost of motoring.

(c) Low cost travel in company cars may reduce the demand for travel by other (more energy efficient?) modes of transport

(d) If company cars are chosen on different criteria from private purchases, engine size, make and model will be influenced. This will have implications for fuel consumption during the whole of the car's life.

Most attention is given here to the consequences (b) and (c) above, since they can give an indication, however uncertain, of the effect of cost on car purchase and use decisions. It is, perhaps, necessary to warn against assuming that a statistical association in the results necessarily means that there is a proven cause and effect mechanism in action. All that can be claimed is that, if there is a strong correlation, and a likely mechanism can be proposed, the results can be regarded as probably cause and effect. This is true for all statistical studies.

In 1978/79, about 31% of the car fleet received some form of monetary support from companies, and these cars accounted for just over 40% of the total car mileage. The mileage driven for different purposes is shown in Table 5.4. Group A and B are company owned cars and accounted for about 8% of the total car fleet: Group C are cars with costs off-set against income tax, which accounted for another 8%: Group D are private cars with mileage related expenses paid: Group E (and Ea - adjusted) are unsubsidised. By weighting the Group A - D cars in accordance with the number in the sample, it can be shown that the average annual total mileage of company cars is 12,200, compared with 7,400 for unsubsidised cars. There are socio-economic grounds for expecting that people with company cars would travel greater distances than those without. Hopkin's best estimate of the difference in non-business mileage between the group receiving maximum subsidy and those with no subsidy, taking into account the effects of socio-economic grouping, was an increase of 1500 - 2000 miles per year. This analysis was not able fully to take account of socio-economic effects and there was wide variability within socio-economic groups, so this finding does not imply that the provision of a company car for the first time, other circumstances remaining unchanged, will increase annual mileage by this proportion. Nevertheless, there is a strong suggestion that company car ownership does encourage greater car use.

TABLE 5.4

Average annual mileage[x], with different types of company assistance

Journey purpose			*Group**				All private cars/
	A	B	C	D	E	Ea	light vans
Business	4,900	6,600	3,100	2,500	300	400	1,100
To/from work	3,400	3,100	1,600	3,000	2,300	2,200	2,200
Essential private	1,300	1,500	1,600	1,600	1,600	1,700	1,600
Discretionary private	3,300	3,600	2,300	2,900	2,700	2,600	2,700
All non-business†	8,700	8,700	6,100	8,000	7,100	7,100	7,000
Other 'business'††	900	200	600	100	0	0	100
Total†	14,500	15,500	9,900	10,500	7,400	7,400	8,300
Number of vehicles	132	175	301	130	2,370	2,370	4,054

x To nearest 100 miles

* Includes all vehicles with complete travel diaries and socio-economic group of main driver known

† Includes 'other' journeys

†† Miles ineligible for the travel diary; these were mainly for travel in the course of work such as delivering goods and journeys by professional drivers but also included journeys away from the public highway.

Note:

Groups A and B are company owned cars.

Group C are cars with costs off-set against income tax.

Group D are private cars with mileage related expenses paid.

Group E (and Ea - adjusted) are unsubsidised.

Source: Hopkin, 1986

TABLE 5.5
New car engine capacity (cc) by subsidy and fuel economy

% of cars	*up to 1000*	*1001-1200*	*1201-1500*	*1501-200*	*over 2000*
Company	7	5	20	59	9
Other	16	12	30	35	7
km/litre	15	12	12	10	7

Source: DVLC, Jan-Nov 1988, plus AA performance data. Quoted by the The Open University, 1989.

Comparing company owned cars (Groups A and B) with unsubsidised cars, Table 5.4 shows that company cars were driven for a higher mileage to and from work, and on discretionary private trips, as well as for business purposes. It is the fuel used for the non-business journeys, including the journey to work, that is of main concern here, though Hopkin gives some evidence that company car holders were less likely than others to travel by rail either on business or to and from work.

The other consequence of company financed motoring is on choice of car. On average, new company-supported cars had engines between 280 and 360cc larger than the unsubsidised. When income and car value were taken into account, company cars were 170cc larger than the unsubsidised. These larger engined cars remain in the car stock for ten years or more, and during most of this time they are used, second hand, in the less subsidised sector. The effects of company car choice on the second hand market has been examined by Mogridge (1985), who concluded that it produced a distortion of the market to larger cars and shorter lifetimes.

Recent data on engine size and fuel consumption have been given by the Open University (1989) and are reproduced in Table 5.5. (They confirm the earlier finding that the company cars had engines about 200cc larger than the unsubsidised.) The same source states that the influence of company cars in 1983 resulted in an additional 3.5 million tonnes of fuel being used (18% of road transport fuel).* However, recent UK Budgets (including March 1991) have increased the tax paid by individuals on company car benefits, so the present effect of company cars on fuel use is likely to be less than this, and should diminish further as the impacts of the tax changes work their way through the national car fleet.

A last consequence is the effect on short term elasticity of car use as influenced by petrol price. An international report (IEA, 1984) found that this elasticity was much lower in the UK than in other countries (USA, Canada, Japan, Germany, Italy). This implies less reduction in car travel as petrol price increases, and was attributed by the IEA to the much higher company car ownership in the UK than in the other countries. Company car holders would be expected to be less sensitive to petrol price increases than unsubsidised motorists, but there are many other factors which affect the fuel price elasticity of car use, including size and density of land use in different countries.

Any proposal to improve the fuel efficiency of the car fleet should not restrict necessary business journeys, but should encourage business, as well as other, travellers to consider more energy

* This large amount presumably includes the excess fuel consumption of company cars during their whole life, through many owners' hands.

efficient vehicles and modes as alternatives. As an example, a company could, as a matter of policy, choose to change to more economical car types. In one case, a company deciding to change to diesel cars found that it was important to make sure that the change applied right up to board member level (Stark, 1983).

The findings discussed above indicate that fiscal measures, and the level of taxation on company cars is one such measure, can play a part in encouraging fuel conservation. Previous Sections have shown how the level of taxation on purchase, ownership, and use can influence car purchase decisions, and the amount of car travel. But this only happens when motorists meet the full costs of their non-business motoring. The present policy of increasing tax on company cars until this is achieved is one way of reaching this goal. An alternative approach would be to pay a flat mileage allowance for business travel, (as for Class D in Table 5.4), an approach which avoids the need for complex tax rules designed to ensure that business travel is paid for, but non-business travel is not. It would be possible to tailor a tax-exempt mileage allowance, like the present Treasury scheme (Inland Revenue, 1990), so that it encouraged fuel economy by restricting it to a flat rate, independent of car model and engine size. It would then be in the user's (financial) interest to use an economical car for business travel, and, of course, also for his non-business journeys. The economical car would also find its way into the second hand fleet (though not perhaps as quickly as the company-owned cars at present which are changed every two or three years), and reinforce and improve the overall fleet fuel economy, instead of making it worse as at present.

It could be, in the future, that companies themselves will see an advantage in moving away from providing cars, with all the administrative and other costs involved, to merely providing a mileage allowance, and letting the owner be responsible for insurance claims and other expenses. A Gallup Poll among company car fleet managers showed already that the rate of accidents is a cause for concern. The average value of claims is also higher for company drivers (at £1,130) than for ordinary motorists (£803) (General Accident, 1990). While it is not certain that the accident rate per mile driven is worse (or better) than ordinary drivers, the higher value claims may reinforce a move towards firms giving mileage allowances rather than cars to their employees.

5.7 Summary

The linkage between road provision and vehicle taxation has a long history, and, while it is not so specific as in the old Road Fund days, rates of taxation for various vehicle classes tend to be kept in line with the costs they impose on the road system.

Taxation of vehicle purchase, ownership and use has been seen to be a means of influencing fuel economy, though there are, of course, many other objectives of taxation. Already , in the UK, more revenue comes from tax and duty on fuel than any other road transport source, but it would be possible to change the balance even further to a "use-based" tax if that is seen to be desirable to conserve fuel. Past experience indicates that direct taxes on fuel may be more effective than indirect ones on engine size (or rated horse-power) in encouraging fuel economy. Company supported business travel by car has been examined, and the lighter taxation of company cars in the past is seen to have distorted the market towards larger cars, shorter lifetimes, and greater fuel consumption.

6. Alternative energy sources for road transport

6.1 Background

A brief introduction to possible alternatives to natural oil for road transport was given at the end of Chapter 2. The main options considered (and summarised schematically again in Figure 6.1) are examined here in rather more detail, though discussion of battery electric road vehicles is reserved for Chapter 7.

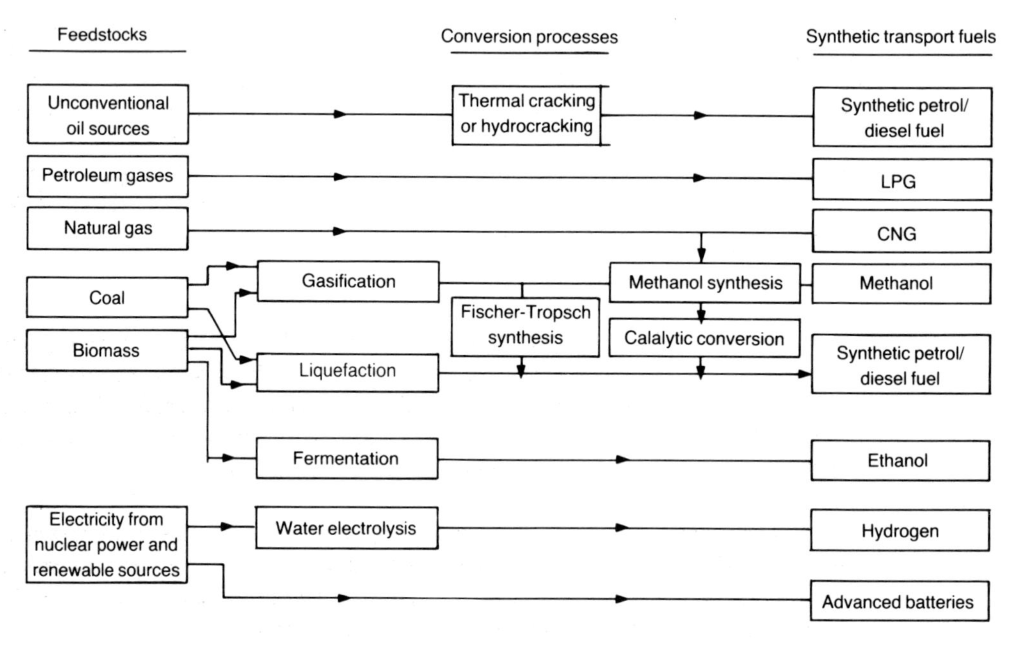

Fig 6.1 *Schematic summary of the main options for synthetic fuels (Source: Langley, 1987)*

Before going into more detail, it is helpful to consider why alternatives to oil are being thought about, and what criteria should be used in assessing their prospects.

In the past, the main incentives for looking at alternatives to natural petroleum were:

(a) The need to have some way of keeping the road transport system running "when the oil ran out".

(b) The value of having a synthetic fuel which could act as a ceiling to restrain the action of oil cartels on oil price.

(c) The wish to have security of supply without the risk of interruptions from the sensitive Middle East (GAO, 1988).

The recent much greater concern about CO_2 emissions and possible global warming has introduced a new dimension into the desirability of different alternative fuels, especially when looking to the longer term. A new criterion is the carbon content of the fuel, or more usefully the CO_2 emission per unit of heat supplied. Some of the basic data for fuels which will be considered in the next paragraphs are collected in Table 6.1.

TABLE 6.1

Specific heat output and carbon dioxide emissions at point of use for various fuels

Fuel	*Chemical formula*	*Gross cal. value MJ/kg*	*CO_2/kg*	*CO_2/MJ*
Coal	70% $CH_{0.5}$	24	2.50	1.104
Petrol/derv	$CH_{1.8}$	43	3.19	0.074
Ethanol	C_2H_5OH	27	1.91	0.071
Methanol	CH_3OH	20	1.38	0.069
Propane (LPG)	C_3H_8	48	3.00	0.063
Natural Gas	CH_4	53	2.75	0.052
Hydrogen	H_2	135	0.00	0.000

Sources: UKAEA, 1989; Kempe's, 1989

The lower CO_2 emission of natural gas (CH_4) compared with petroleum and especially coal are very evident, though it must be stressed that the figures given in Table 6.1 are for emissions at point of use: emissions from refinery or plant are not included. Ethanol, methanol and also propane do not seem to have so large an advantage over petroleum that the case for using them necessarily overcomes the difficulties associated with any change of fuel.

The CO_2 emissions of different fuels used in power stations are shown in Table 6.2. This shows the reduction that can be gained by switching from coal to natural gas, and will be referred to later when assessing the potential of electric road vehicles. Both Tables act as a useful additional framework when discussing alternatives for road transport.

However, the cost of the alternative fuel remains an important factor in assessing its prospects, and it is generally accepted that none of the possible alternative fuels can compete in cost with products from natural crude oil, until scarcity or long-term supply restrictions push the price up to two or three times present levels (say, to $50 to $70 per barrel).

In the following paragraphs, therefore, the main criteria used are the estimated cost and, for longer term considerations, the effect on CO_2 emissions.

TABLE 6.2
Relative carbon dioxide produced from power stations burning different fuels, and with various generation cycles

Plant Type	Efficiency (%)	Grams of carbon per Mega-Joule (Electricity)	Ratio of CO_2 produced to CO_2 from current UK Power Stations
Coal			
Current coal fired	31.5	76.2	1.00
New coal fired	37.0	64.9	0.85
Gasification combined cycle using advanced gas turbines	40.0	61.0	0.80
Natural Gas			
Modern combined cycle gas turbines	42.0	33.3	0.44
Advanced combined cycle gas turbines	47.0	29.8	0.39
Predicted combined cycle gas turbine	50.0	28.0	0.37

Source: British Gas Plc, 1989

6.2 Liquids from coal

The main option that has been studied at TRRL is the conversion of coal to liquid fuel suitable for conventional petrol or diesel engines. Studies were based on the processes then being developed by the National Coal Board (1978). National and world-wide reserves of coal are very large when compared with proved reserves of oil (BP, 1989), and they are geographically more widely spread. The basic feedstock could meet demands for hundreds of years, rather than tens of years for oil. Early work at TRRL (Welsby, 1974) saw liquids from coal as a long term alternative to natural oil in function, and as a price ceiling which could protect the user from excessive monopoly prices for natural oil. Other studies (Chapman *et al*, 1976) felt that coal should be converted to gas and used in small combined heat and power installations, leaving road transport to be based on battery powered electric vehicles. This conclusion was not entirely accepted (Porter and Fitchie, 1977) even though, as Figure 6.2 shows, a car with a sodium sulphur battery used less energy. A car fuelled by synthetic gasoline had lower total costs, as well as the advantage of unrestricted range, and these factors were thought to outweigh the energy comparison.

However, this research did not explicitly take account of the relative emissions of CO_2 from the whole chain of processes from feedstock fuel conversion to use on the vehicle. Porter (1979) recognised that liquids from coal were not the only solution to depletion of natural oil. He noted that the burning of fossil fuels might lead to a risk of climatic change from the increased concentration of CO_2. The use of electric battery cars, re-fuelled by nuclear electricity, would alleviate this problem, but introduce other environmental and safety factors. From the perspective of the 1990's, the nuclear power option seems to many people rather less attractive than it did in 1979.

In a 1978 review of long term possibilities for energy for road transport, the Advisory Council on Energy Conservation concluded that while coal was available, liquids derived from it were

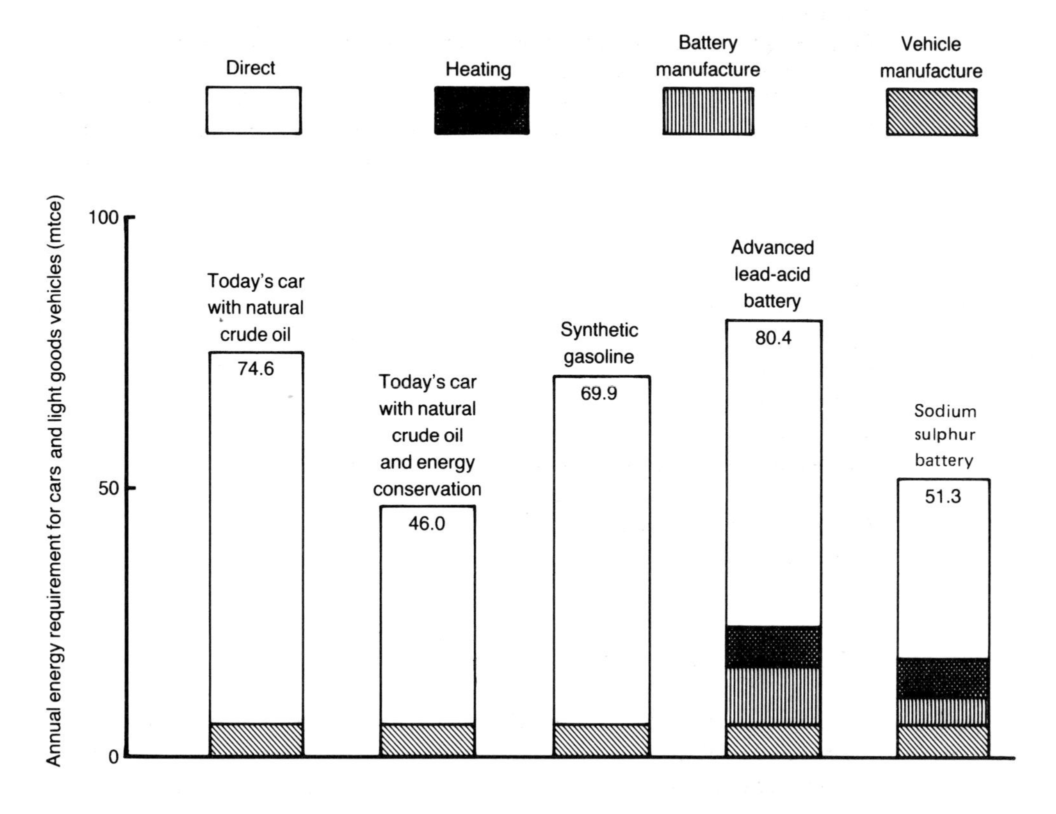

Fig 6.2 *Annual energy requirements for cars and light vehicles in 2025 (Source: Porter and Fitchie, 1977)*

the most likely choice (ACEC, 1978). In a more recent review, Langley (1987) still concluded that petrol from coal by direct liquefaction was a front runner in the UK was a front runner in the UK (second to petrol from residual fuel oil or heavy crude). The various alternative fuels considered by Langley were shown in Figure 6.1, and a summary of the synthetic fuel economics (at 1984 prices) is given in Table 6.3. The sensitivity of liquid fuel cost to feedstock (coal) cost and the efficiency of various processes is shown diagrammatically in Figure 6.3. On economic grounds, liquids from coal were not seen as likely to be viable until oil prices reached around $55/barrel in 1985 prices (Davies, 1985). With process efficiencies of the order of 60% or less (energy in final fuel compared with energy in feedstock), increases in CO_2 emissions in total seem inevitable by comparison with natural oil. The great advantage of the option is, of course, that the vehicle fleet can make fairly minor changes to its technology, and can adapt slowly to the new fuel. But with the present concern over greenhouse gases, the use of coal as a starting point looks less of a first choice than it did a few years ago.

What seems to be left out from these analyses is the additional CO_2 emissions due to the whole conversion process, so that they can be added to the CO_2 emissions at point of use (as shown for some fuels in Table 6.1). The techniques of energy analysis (Chapman *et al*, 1974, for example) could be adapted to this new requirement, and might possibly revise the conclusions about the ranking order of synthetic fuels for road transport.

TABLE 6.3
Summary of synthetic fuel economics at 1984 prices

Assumptions		*Original Units*	*$/GJ*
Feedstock prices:	coal	45 £/tonne	2.33
	natural gas	30 p/therm	3.70
	electricity	3 p/kWh	10.83
	crude oil	28 $/barrel	4.57
	fuel oil/very heavy crude	90% of crude	4.12
	LPG	35 p/therm	4.31
Economic parameters:	discount rate	10%	
	exchange rate	1.40 $/£	
	currency basis	Jan 1984 $	
Product Costs			*$/GJ*
Petrol from fuel oil/very heavy crude			6.00
Petrol from LPG (Cyclar)			7.50 (d)
Methanol from natural gas (ICI)			8.40
Petrol from coal (2-Stage)			10.20 (d)
Petrol from natural gas (Mobil)			10.60
Kerosene/gasoil from coal (Shell)			11.00 (d)
Petrol from coal (H-Coal)			12.00 (d)
Petrol from coal (EDS)			12.10 (d)
Methanol from coal (Lurgi)			14.40
Methanol from wood			14.70 (d)
Ethanol from sugar beet			15.40 (d)
Petrol from coal (Lurgi/Mobil)			18.00 (d)
Petrol from coal (Fischer-Tropsch-SASOL II)			19.10
Liquid Hydrogen from Electrolysis			19.70
Petrol from coal (Fischer-Tropsch-SASOL I)			19.70

(d) under development: estimated costs only
Note: Hydrogen gas from electrolysis would cost about $6 - $7/GJ less than the liquid form.
Source: Langley, 1987

6.3 Other alternative fuels

If liquids from coal have become less attractive, are there other fuels which would substantially reduce CO_2 emissions, without offsetting disadvantages? The Sixth Report of the Energy Committee (House of Commons, 1989 I) quoted the potential for reducing emissions per vehicle as set out in Table 6.4. (The Table appears to refer to CO_2 emissions only.)

The figures in the Table need some amplification. The estimated reduction in emissions from changing to diesel vehicles is rather more than seems likely from the estimates given in Chapter 3, where a 15% to 20% reduction in fuel consumption is more what is now expected when compared with modern petrol cars. The reduction from using gas in vehicles (presumably LPG or CNG) stems from the lower carbon content of the fuel, though the gain is larger than would have been expected from the figures in Table 6.1, and there would be disadvantages in extra weight and bulk of fuel tanks. The figure for bio-alcohol probably refers to ethanol, obtained from fermentation of natural grown material. While the fermentation process itself produces CO_2, this

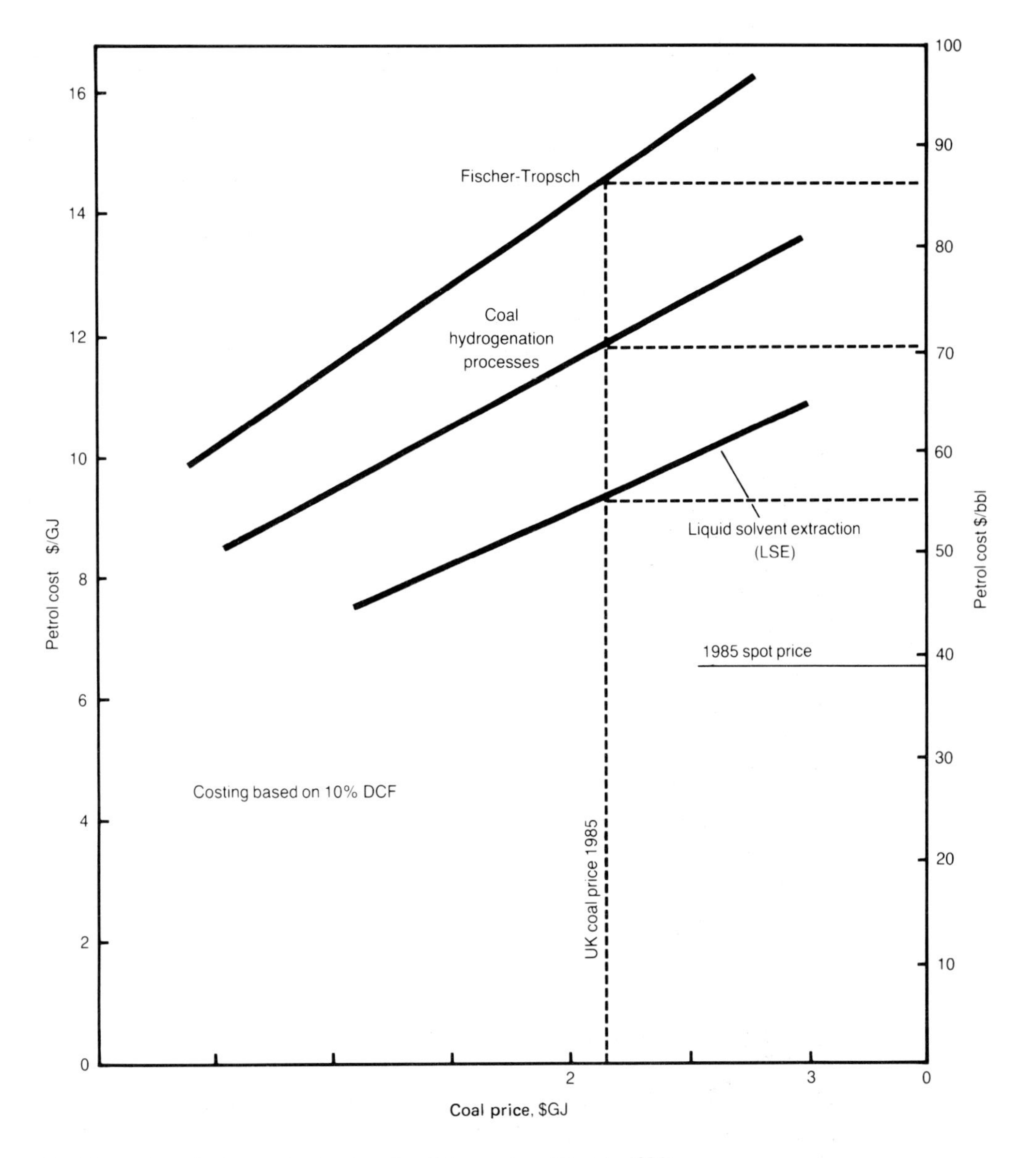

Fig 6.3 *Liquefaction economics (based on IEA cost of coal liquids, 1984)*
(Source: Davies, 1985)

can be considered as part of the bio-mass cycle, rather than an addition from fossil fuel use. Methanol is considered in more detail in the next Section, but it has to be obtained from distillation of wood or fossil fuel: it thus may have penalties in energy and in CO_2 emissions, though if the wood is obtained from sustainable growth in forests the process can be considered as part of global carbon re-cycling.

TABLE 6.4
Potential for reducing emissions (CO_2) from road transport

Action	*Resulting reduced emissions per vehicle*	
Diesel for petrol	up to 30 per cent	
Gas for all vehicles	up to 45 per cent	
Bio-alcohol for cars	100 per cent	
Hydrogen for all vehicles	100 per cent)	if electricity from non fossil sources
Electric vehicles	100 per cent)	

Source: House of Commons, 1989 I, quoting ETSU's paper to the No 10 Seminar, April 1989.

Table 6.5 summarises the advantages and disadvantages of gas, bio-mass, and methanol alternatives. There are consequences, particularly with the latter, for engine and fuel system design, but many experimental vehicles have been run in a number of countries on these fuels, and cars running on Liquid Petroleum Gas are common in the Netherlands. Brazil has had a major programme to promote the use of ethanol fuel in cars. There the objective was not CO_2 reduction, but rather to gain independence from imported oil. Urban buses running on compressed natural

TABLE 6.5
Advantages and disadvantages of some alternative fuels

PROPANE (C_3H_8) (LPG - LIQUIFIED PETROLEUM GAS) *Advantages* Low exhaust emissions; high octane number; liquefied at moderate pressures (160psi @ 70°F) *Disadvantages* Power loss; poor cetane number; narrow flammability limits *Sources* Crude-oil wells; natural gas wells; by-product of refinery operations	METHANE (CH_4) (NATURAL GAS WITH MORE THAN 84% METHANE) *Advantages* Low exhaust emissions, low particulates; low reactivity in the atmosphere; high octane number; density less than air (dissipates quickly); good cold start *Disadvantages* Low boiling point (-258.5°F); poor energy density on volumetric basis (limited range); power loss; poor cetane number *Sources* Natural gas wells (New Zealand, Australia, Mexico, Canada); anaerobic digestion (methane bacteria); coal
METHANOL (CH_3OH) *Advantages* High octane number; low flame temperature (lower NO_x, low deposits); broad flammability limits (smooth operation in fuel-lean region); high latent heat of vaporisation (increase in volumetric efficiency, decrease in work of compression); more moles of product per mole of reactant *Disadvantages* Poor cetane number; poor energy content on volume basis; high latent heat of vaporisation (more difficult cold start); polarity (materials compatibility, corrosion); aldehyde emissions *Sources* Coal; wood; natural gas; any organic waste	ETHANOL (C_2H_5OH) *Advantages* Much the same as methanol (increase in efficiency and power somewhat less) *Disadvantages* Much the same as methanol (energy content somewhat better); cost; impact on food production *Sources* Fermentation of organic matter (grains, sugar cane, sweet sorghum); ethylene (petroleum derived feedstock)

Source: Ford,1981

gas have been ordered for the Australian city of Adelaide, with the objective of achieving low noise and low exhaust emission levels (MAN, 1991).

Recent reviews of alternative fuels include one on bio-ethanol as a source of energy, which has been carried out for the Department of Energy, and a review of alternative transport fuels by the International Energy Agency (Department of Energy, 1989a).

6.4 Methanol

The possible use of methanol as an alternative to petrol was given a boost in the USA by the proposed amendments to the Clean Air Act (Gavaghan, 1989). One element is the call to have sold 500,000 "clean fuel technology cars" by 1995, and, although the Bill defines "clean fuel" as methanol, ethanol, propanol, or electricity, the generally accepted view is that methanol is likely to be the most practical way of meeting the target.

The main pro's and con's for methanol have been outlined in Table 6.5. Disadvantages, at the point of use in the vehicle, include low energy content on a volume basis, more difficulty with cold starting, low cetane number (so it is not a direct substitute for diesel fuel), and corrosion problems with some common materials in car fuel systems. Exhaust emissions contain aldehydes, which are carcinogens. But there are also advantages, including a high octane number, and enhanced lean-burn capability because of broad flammability limits, which can improve engine efficiency and allow the use of a smaller engine for the same road performance. A description of the development by Volkswagen of a pure methanol car was given by Menrad *et al.* (1977).

More recently the case for the methanol car has been put persuasively by Gray and Alson (1989), from the American viewpoint. Some of the proposals, like the use of a system to stop the engine when the car slows down are not special to a methanol car, and could be used to save fuel on a conventional vehicle. (The VW Formel E models used a similar device in production cars some years ago - a technical description is given by Neumann, 1989.) The approach by Gray and Alson to the problem of feedstock for the methanol fuel, and dealing with CO_2 emissions at the plant by liquefaction, seem rather optimistic. But the pressure in the USA on improving air quality, particularly in California, may well press the claims of methanol more vigorously than under European conditions. Again, the effect of moving to methanol on total CO_2 emissions may merit further study, but the advantage of the fuel in terms of CO_2/MJ compared with petroleum is only about 7% at the point of use (Table 6.1), so it is not by any means a "solution" to the CO_2 problem.*

6.5 Hydrogen as a transport fuel

Hydrogen has obvious attractions as a non-polluting fuel because its products of combustion contain no carbon compounds. The Select Committee on Science and Technology (House of Lords, 1989 I) noted that hydrogen can be stored and distributed through pipelines, and so the

* Unless the methanol is produced from sustainable sources like crops or managed soft-wood forests - which seems unlikely.

"hydrogen economy" has great attractions, provided that the electricity to make the hydrogen is not produced by burning fossil fuels. Apart from the so-called renewable resources (Department of Energy, 1988), there are two suitable sources of electricity: nuclear power and photo-voltaic solar cells. It was stated by Archer (1989) that there is no spare nuclear capacity in the UK at present (though this may not be true of other countries like France). And while there are "prospects of dramatic advances in thin-film solar cell technology", some authorities are sceptical about such claims (House of Lords, 1989 I). The medium term (early next century) chances for substantial non-fossil fuel generation of hydrogen cannot be viewed with optimism.

Hydrogen can also be produced, along with carbon monoxide, by the partial combustion of coal in the presence of steam. However, this is a fairly inefficient process which would almost certainly not lead to reduced overall emissions of CO_2, though again a critical review based on energy analysis might be justified.

While hydrogen has attractions as a gaseous fuel, its use in road transport is far from easy. If it is stored as a compressed gas, the weight and bulk of the containers are a considerable penalty. Hydrogen in liquid form has to be stored at around 20°K (Absolute), which carries very severe penalties in weight, cost and safety for ground transport. A more promising possibility is storage in the form of metal hydrides, which are metal alloys able to store the hydrogen between their molecules, and release the gas when heated. Hydrides have been used as hydrogen "stores" in experimental cars and buses (Billings, 1976), and some of the practical problems in storage and release of hydrogen, and in the necessary modifications to a car engine, have been described by Watson *et al*, (1984). Recent experimental work in Japan with a hydride storage system in a road vehicle has been described by Hama *et al* (1988). The problems are not insuperable, given a supply of hydrogen, but there are likely to be serious weight and cost penalties.

It will be clear from the previous paragraphs that the use of hydrogen as a transport fuel is a very long term possibility. Some authorities put it more strongly. The Department of Energy (1989b) reviewed the situation since their last study was published (Langley, 1983), and concluded that "only under the most extreme CO_2 emission control and energy scenarios would the 'hydrogen economy' look reasonable.... A significant move towards hydrogen, even if restricted to the transport sector, would involve enormous infra-structure costs, and, depending on the source of the gas, probably huge economic and environmental costs as well".

6.6 Summary

In the light of falling real costs of natural crude oil, and greater importance attached to reducing CO_2 emissions, the attraction of liquid fuel from coal appears to be reduced compared with the prospects 5 to 10 years ago.

Other alternatives, like methanol, may be viable in conditions of oil scarcity, but there are implications for engine design which make the introduction of a new fuel expensive, and they do not appear to reduce CO_2 emissions sufficiently (compared with petrol or diesel fuel) to make their introduction imperative. Special requirements for exhaust emission control to improve air quality may press the claims for methanol use in the USA.

The use of hydrogen as a transport fuel can only be considered as a very long term possibility.

7. Electric road vehicles: An alternative?

7.1 Background

The main subject for this Chapter is the potential of the battery electric road vehicle for providing a practical alternative to oil-burning internal combustion engined vehicles in road transport. The main attractions have always been the ability to use electricity generated from many alternative sources, including non-fossil fuel, and the absence of exhaust emissions from the vehicle (though power station emissions have to be considered). Disadvantages have been the limited driving distance (range) on one battery charge, extra cost, especially when replacement batteries are included, and a "milk float" image of low performance. This last disadvantage is unfair, for modern electric urban cars and vans, as will be seen, can have road performance fully compatible with other traffic. But the criticism of limited range is more valid. One commentator put it succinctly: "who would buy an expensive small car with a one gallon tank which took eight hours to refill?"

While this sort of comment is crucial to the possible replacement of all internal combustion engined vehicles by battery electric ones, it does not necessarily mean that a limited range vehicle has no practical use. There are applications, particularly in large cities, where a daily range of around 100km can be adequate. Although the battery electric milk float is bad for the "image" of modern electric vehicles, its characteristics and economics have, uniquely in the UK, allowed daily milk delivery to homes to continue into the 1990's. There is a niche for even a very limited performance vehicle. The question is, are there other suitable applications?

Going back to the early history of the motor vehicle, it appeared to some enthusiasts in the early 1900's that a battery electric urban car had considerable positive advantages over the (then) unreliable petrol or steam car. Joel (1903) is the source of the often-quoted remark that, with more than 200,000 horses stabled in London overnight, there was a need for the daily removal of at least 5000 tons of manure and refuse, "in addition to what is distributed over the roads or finds its way into the residences of the people". He looked to the substitute of motor-cars for horse to relieve this unsanitary state, and put forward a detailed case for the battery electric motor car of the day. Among the advantages claimed were noiseless traction (with neither smell nor refuse); easy (engine) starting; and features such as regenerative braking, not available for other power sources. The daily range was enough for most town applications, and the performance (particularly braking) better than horse-drawn transport or other contemporary motor cars. Even the difficulty of obtaining a fresh electrical charge was being overcome, "by the readiness of the electric-supply companies in towns and country districts to provide motor-houses, and to recharge at all hours of the day or night, at reasonable cost". In the discussion on Joel's paper, much was made of the high cost of keeping a horse-drawn carriage and driver: it was thought that the electric car was so simple that "the car could be driven by the ladies themselves, thereby saving the driver's wages".

TABLE 7.1
Batteries for road transport: values used for assessment in the early 1970s

Battery Type	*Energy Density (kWh/tonne)*	*Cycle life*	*Capital Cost (£/kWh)*
Lead/Acid traction (Tubular plates)	20	2000	30
Improved Lead/Acid (Flat plate)	50	500 - 1000	10
Metal/Air (Iron/Air)	100	500 - 1000	25
High Temperature (Sodium Sulphur - operating at 300°C to 400°C)	200	500 - 1000	5
(Petrol - used at 20% efficiency)	2,500	-	-

Source: Waters and Porter, 1974

Unfortunately, this optimistic scenario did not come about. Partly this was due to the rapid development, in power and reliability, of the internal combustion engine and its vehicle. This development, which also eliminated the steam car, contrasts with the very slow development of the electric car, and in particular the storage battery.

The key part played by battery development can be illustrated by looking at the battery performance values used for assessment purposes in the early 1970's, and then in the early 1980's. In the earlier study (Waters and Porter, 1974), the base line battery for comparison was the conventional lead/acid traction battery, with tubular plates, and designed for a long life of deep discharge cycles.*

For comparison with the traction battery, three other types were considered promising: the improved lead/acid battery (developed from the car SLI battery): the metal/air battery (at that time, the iron/air couple showed promise): the high temperature battery (based on the sodium/ sulphur cell).

The performance values used for assessment in the early 1970's are shown in Table 7.1. The high cost and weight, but long life, of the lead/acid traction battery contrasts with the expected improvements in lead/acid (though with shorter cycle life), and the notable prize held out by the development, then in progress, of the high temperature sodium sulphur battery.

It is worth stressing the importance, in operating cost terms, of a high cycle life. The first cost of a battery pack for a road vehicle may be several thousand pounds. The replacement cost at the end of its cycle life is therefore an important element in the cost comparison of battery electric and internal combustion engined vehicle.

The situation on battery development 10 years later is shown in Table 7.2 (Gyenes, 1984). The favoured batteries for development are now nickel-zinc (instead of iron/air), but still advanced lead/acid and the sodium sulphur high temperature battery. Ten years of development have

* This deep discharge life is, incidentally, one characteristic that is essential for transport use, and is not present in the duty of the more usual "starter, lights and ignition" (SLI) car battery. It has proved difficult to develop the much lighter car battery to be suitable for traction use.

TABLE 7.2
Characteristics of batteries for road transport applications

	Lead-acid Tubular	*Lead-acid Flat Plate*	*Advanced Lead-acid*	*Nickel-zinc*	*Sodium-sulphur*
Power density* (kW/tonne)	42	51	66	130	130
Energy density** (kWh/tonne)	25	33	44	92	120
Energy density** (kWh/m^3)	65	90	107	162	135
Charge-discharge efficiency** (%)	75	70	75	75	75
Cycle life***	1500	500	500	300	1500
Capital cost (£/kWh)	100	45	45	45	45
Possible date for production	1980-85	1980-85	1985-90+	1990-2000	1990-2000

*Short duration peak power density at 80% depth of discharge. Half of this power is available for at least 1 hour from 0 to 80% depth of discharge.
**3 hour discharge rate from 0 to 80% depth of discharge.
***Battery life in years is (Cycle Life x Working Range)/Annual kms of Travel.
+Not yet in production
Source: Gyenes, 1984

modified the hopes for the batteries under development. Both cycle life and energy density estimates have been reduced for the advanced lead acid battery. For the sodium sulphur, cycle life is now thought to be higher, but the energy density is considerably less, as the need for thermal control and insulation of the battery pack has become evident.

These two sets of figures are not intended to denigrate the considerable research and development effort that has gone into batteries for transport purposes, but rather to show the difficulties of producing a suitable battery, and to serve as a warning against too much optimism when new batteries appear to be "just around the corner".*

This review does not attempt to go into the electro-chemistry of batteries. If more information is needed, the excellent book edited by M Barak (1980) on battery fundamentals should be consulted. Particular batteries which are being (or have been) under development for road transport purposes are described in references which are quoted later in this Chapter.

Having set the scene for battery electric vehicles as an element in using an energy source other than oil for transport, the next Sections of this Chapter consider first the short to medium team developments which have been carried out, and then look at some of the long-term prospects.

* See, for example, present interest in a capacitor type battery under development by Isuzo and Fuiji Electrochemical in Japan, for which claims of very high energy density and rapid re-charging are made (Woodward, 1991).

7.2 Short to medium term developments

It has been seen that the battery electric road vehicle was defeated by the internal combustion engine in the first decade of this century. Its revival has long been on the horizon. In the early 1970's, prospects were enhanced by growing awareness of the noise and fumes from conventional traffic, and a review of market prospects foresaw some penetration of electric light goods vehicles and urban buses into the conventional fleet (Waters and Porter, 1974). This assessment was based on equivalent annual costs, with a penalty placed on the internal combustion engined vehicles to reduce noise and emissions. It also assumed that improved lead acid batteries would be available, and that the development of the advanced batteries would be successful.* (The assumptions about battery performance have been set out in Table 7.1.)

The short-term prospects for urban buses and light vans (but not for private cars) looked good enough in commercial terms for industry to respond to (financial) encouragement from the Department of Industry, and to start the development of prototype urban buses, and a small scale production of urban delivery vehicles. Trials of a 26 passenger urban bus (Saunders, 1976) showed that, while operators were sympathetic towards an electric bus, they required it to be as reliable and cheap to operate as the equivalent diesel. Not surprisingly, in view of its prototype stage, the bus could not demonstrate either quality. Other buses, notably the Chloride "Silent Rider" with a payload of 50 - 60 passengers, were built and operated by bus companies in an experimental way (Harding and Morris, 1974), but were not able to compete in operating cost.

It may be that, in the absence of dramatic improvements in battery performance, a return to some form of the trolley bus would be more viable in some urban areas. The feasibility and economics of a hybrid battery/trolley bus, able to operate for a limited distance away from the power lines, was examined by Weeks (1978a), but was not taken further in the UK. However, experimental schemes were tried in Germany, and a conference in 1981 heard several papers describing developments and proposals (EVDG, 1981). More recently, an international study of hybrid trolley buses (COST 303, 1987) has reviewed their technical development and economic potential. In this country there is interest in re-introducing conventional trolley buses in Bradford, the last city to abandon them in 1972. (They are, of course, common on the Continent, as are light rail transit and trams.)

The development and demonstration of urban delivery vans with 1 tonne and 1.75 tonne payload was more successful (see, for example, Harding, 1977.) With the electric version of the Bedford CF 1 tonne payload van, a detachable and ventilated battery pack was developed so that a discharged set of batteries could be replaced, within minutes, by a fully charged replacement. The dangers (from hydrogen explosion) when recharging and topping the batteries up with distilled water were recognised, and operational procedures laid down for greater safety (Sims, 1984).

In November 1977, Lucas, Chloride, and Crompton Electricars, with the Department of Industry and in association with the Greater London Council, sponsored a major demonstration: 66 battery electric delivery vans of three types were used by operators, and 37 internal combustion engined vans provided "control" data for comparison. The London Electric Delivery Van Assessment

---

* It is, incidentally, assumed in this Section that electric vehicles will have a road performance close to that of conventional ones: they are not "milk float" types.

Scheme was described by Wicken (1979), and some preliminary results published (Wicken and Murray, 1980). The trial lasted for 5 years, and TRRL was responsible for monitoring and analysing the results for the Department of Trade and Industry; the conclusions have not been published. In a separate trial, up to 73 of the same vehicle types have been operated by the Southern Electricity Board since 1978, and some results have been published for 3,000,000 km running in service. It has been concluded that maintenance costs for the electric vehicles were 30% less than for the internal combustion engined vehicles, even after allowing for extra costs due to the time taken for topping up the battery water after each re-charge (quoted by Fabre, *et al*, 1987).

The above trials were mainly intended to prove (or to disprove) the commercial viability of state-of-the-art battery electric buses and delivery vans. A comparative trial of the energy consumption of an electric, petrol, and diesel urban van was reported by Wood *et al* (1981), and by Waters and Laker (1982). It is, of course, difficult to find a fair comparison between the energy used by vehicles powered by electricity and by petroleum, but the method adopted was to compare on the basis of crude oil, either refined to petrol or derv for the internal combustion engined vehicles, or to fuel oil and used in power stations to produce electricity. (Wood *et al* also showed that the primary energy used by the electric vehicle was the same if the electricity was generated by coal burning.)

On this basis, the primary energy used, plotted against section average speed in central London traffic, is shown in Figure 7.1. The electric van used more primary energy than the petrol van, except in the most congested traffic: the diesel was appreciably more economical with energy than either. These results, and the way they were presented, caused considerable controversy. As only about 12% of electricity was generated from oil at the time, the electric vehicle manufacturers felt that it was unfair not to point out, much more strongly, that the electric vehicle could, in effect, use coal or nuclear produced electricity, and that this flexibility of fuel supply was an advantage.

When the relative CO_2 emissions are taken into account, the picture changes again. Using the data in Tables 6.1 and 6.2, it is possible to make an approximate adjustment to the energy curves of Figure 7.1. Emissions of CO_2 per kWh of electricity produced are roughly in the ratio 100:80:40 for coal, oil, and the advanced use of natural gas. Figure 7.2 therefore shows the relative effect of different kinds of electricity generation on CO_2 emissions for the electric vehicle. With coal burning, as might be expected, emissions are higher than either the petrol or diesel. With oil, the electric vehicle has an advantage over the petrol van in very congested conditions, But with advanced natural gas generation, the electric vehicle matches the low emissions of the diesel van. The same result could be obtained if the electricity were generated 50% by nuclear plant, and 50% by coal burning. The conclusion is that the battery electric road vehicle may reduce CO_2 emissions, compared with petrol vans, but will only be equivalent to the diesel van if the appropriate mix of fuels in electricity generation is available.

The most thorough recent review of the technical and economic conditions likely to favour the use of battery electric road vehicles was carried out by a European study group, COST 302 (Fabre *et al*, 1987). Much attention was given to the possible impact in Europe of electric vehicles on energy and on the environment (air quality and noise), though their influence on CO_2 emissions was not covered. Care was taken in the study to compare present designs of electric vehicle with their internal combustion engined counterparts, both in terms of their resource costs (to the

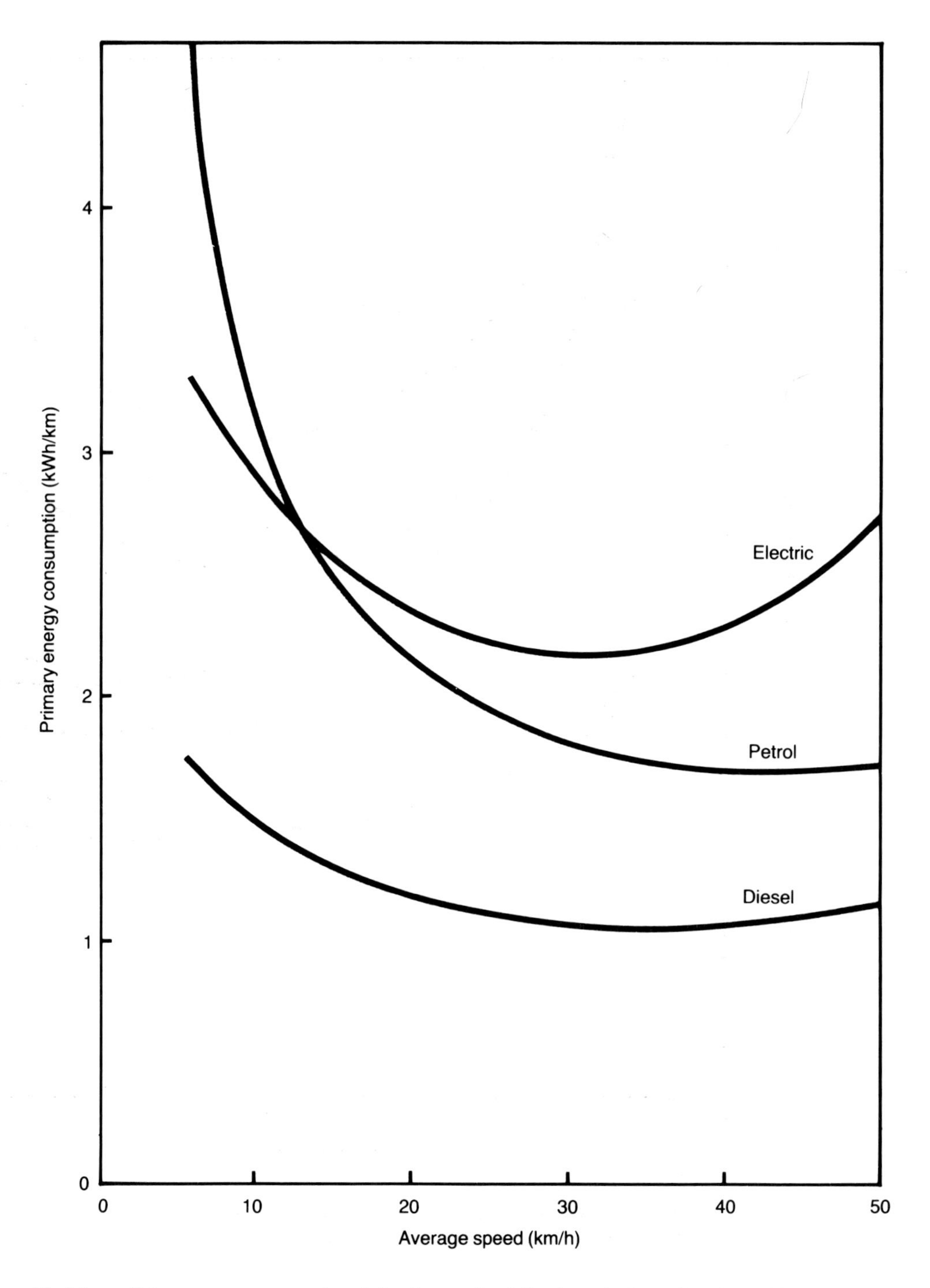

Fig 7.1 *Primary energy consumption of electric, petrol and diesel vans in Central London traffic (Source: Wood et al, 1981)*

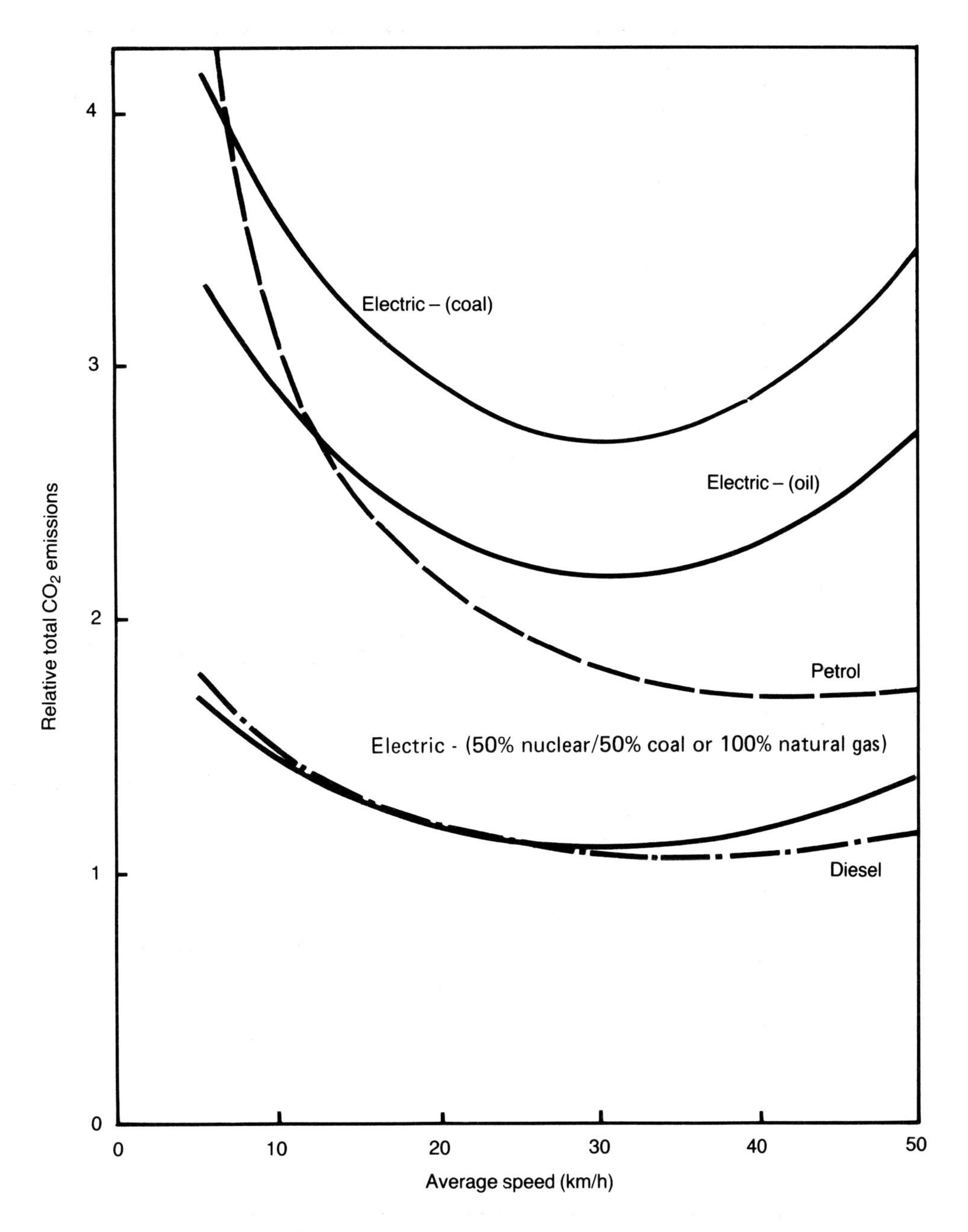

Fig 7.2 *Relative carbon dioxide emissions of electric, petrol and diesel vans (Source: Wood et al, 1981 and Tables 6.1 and 6.2)*

community as a whole, and thus excluding taxation), and in terms of their commercial costs to operators and private motorists. The main conclusions were that present day electric vehicles could, in function, be suitable to replace 6 million small cars (7% of the fleet) and 1 million light vans (12%) in Western Europe, but this would only happen if they could be made and sold much more cheaply than present day electric vehicles. There would be a saving of about 3.5% of total transport fuel, because of substitution by non-fossil fuel electricity, and little difficulty was seen in providing electricity for re-charging through the existing supply system.

The assumed penetration of battery electric road vehicles in Europe was estimated to reduce air pollution from CO, NO_x, and HC by as much as 20% to 30% in the urban areas where they would be likely to be used. Overall, the reduction of emissions would be very much less, more in line with the 3.5% reduction of transport fuel. In Great Britain and Germany there would be some additional emission of sulphur dioxide from coal fired power stations. The report concluded that lower whole life costs of batteries was necessary, and that some of the advanced battery systems, available at laboratory scale at present, have promise in this direction. It was stressed that, in future, battery research should focus on industrial production and cost reduction. The general tone of the study was hopeful, but again pointed to the vital need for improved batteries for electric vehicles.

7.3 Long term prospects

It is clear from the development of electric vehicles based on present lead acid batteries that their prospects would be considerably improved if a battery appeared which had:

- High energy/power density.
- Long life with deep discharge cycles.
- Low cost per unit of stored energy.
- Rapid recharge capability.

Many candidates for battery development have, as has been seen, potential and promise, but so far none has demonstrated all four qualities in a production battery system. The particular problems of traction battery design were touched on briefly in the introductory paragraph, and here it is merely reiterated that most assessments of battery electric vehicles for the future make assumptions about the existence of improved batteries of the following general types:

1. Advanced Lead Acid. Like today's traction battery, but with the low cost and high energy density of car batteries, without their low cycle life.

2. Nickel/Zinc. An alternative low temperature battery, typical of several presently under development.

3. Sodium Sulphur. A battery which functions at 300°C to 400°C, and potentially has excellent characteristics for traction purposes.

Useful short descriptions of the development of advanced batteries are given by Sims (1984) for the lead acid type, and by Dell (1984) for high temperature and other batteries.

Use of these batteries, when available, could make a whole range of battery electric road vehicles more attractive in performance and, eventually as oil becomes more expensive, in cost terms, as well. Gyenes (1984) gives a consistent design evaluation of vehicles based on these future batteries (Table 7.2), and shows that even a 20 tonne payload HGV appears feasible if the sodium sulphur battery development is successful. Small battery cars and urban vans of (probably) acceptable road performance appear practicable with advanced lead acid batteries, but are not likely to be attractive in cost until the real price of liquid fuel has doubled.

The restricted range of battery electric road vehicles before re-charging, which is particularly inhibiting for private cars, has led to a number of studies of the need for, and the consequences of, an infra-structure for battery exchange or re-charging points (Weeks, 1978b and Watson *et al*, 1986). It was concluded that the re-fuelling infra-structure could be provided without excessive additional cost, especially if re-charging of batteries overnight at home were a possibility for most users. This assumes that the majority of motorists park their cars at home off the street, and, in a very extensive study, Watson *et al* consider the implications for those who do not.

Another possibility for increasing range is the hybrid vehicle - a combination of battery and small internal combustion engine which could give unlimited range (see, for example, Adams, 1979 and Mitcham, 1979). There are a number of ways that the technology can be put together, and while there are advantages to be gained, sometimes almost as a bonus (for example, by saving braking energy in regenerative systems), there are also some fears that the complexity of the vehicle propulsion system, and its cost, may lead to the "worst of both worlds" being realised instead of the best. There was a period in the early 1980's when many hybrid urban vehicles were being developed and run experimentally (see, for example, papers in EVDG, 1981, and University of Strathclyde, 1984). Experimental work on hybrid electric vehicles slackened off in the mid-1980's, though a paper by Brusaglino and Mazzon (1989) gives an account of developments by Fiat in Italy, and present work in INRETS in France (not yet published) may indicate that interest is reviving. One cause of revival is the drive towards improved air quality in the Los Angeles basin in California. One of three vehicles sponsored by the Los Angeles City Council and the city's Department of Water and Power is the hybrid LA301 which is reported (Autocar and Motor, 1990a) to have a less than 1 litre internal combustion engine and a 45 bhp electric motor powered by sealed lead acid batteries. The four seater has enough power from the batteries alone to drive at just under 70 mph for a range of 60 miles. With the internal combustion operating in drive/charge mode, the range can be extended to 125 miles, while still meeting the Ultra Low Emission Vehicle (ULEV) regulations which give incentives to owners and operators of the cars - in purchase price, cheap electricity for re-charging and grants for fleet buyers. Altogether, the pressure to improve air quality in Los Angeles may encourage more rapid development of small hybrid cars. They have advantages in fuel economy and emissions because their internal combustion engines only have to be run at optimum speed and power, instead of operating over a wide range where it is more difficult to reduce emissions and fuel use.

In contrast, the announcements in the technical press in 1990 about the marketing of electric cars and light vans (by Fiat and General Motors) seem to be describing present day state-of-the-art lead acid battery vehicles. As the Fiat Panda Elettra car is said to be priced at twice the price of the petrol version - which has four rather than two seats - it can only expect to be sold for use in areas, like environmentally sensitive city centres, where the internal combustion engine is banned. (Incidentally, the Enfield car, sponsored by the Electricity Council (1980), was a vehicle of a very similar kind, and about 70 were manufactured in 1966. Progress with battery electric road vehicles often seems to be painfully slow.)

However, the GM Impact electric car is much more advanced in concept, though it is still only a two-seater, and uses lead-acid batteries (Autocar and Motor, 1990b). Priority has been given by the designers to high acceleration and top speed. It has been shown to out-accelerate a conventional Vauxhall Cavalier, and is claimed to have a maximum speed of 100 mph. A range on a single charge of 120 miles is predicted. The main technical innovation is in the use of AC motors, with solid state converters from the DC batteries. GM admit that cost is still a problem, as is the replacement cost of a battery pack (£1000) every 20,000 miles. But the car is seen as a future possible contender in the US second car market for short journeys and commuting, and could presumably be used in Los Angeles under the ULEV regulations.

More than ten years ago, the medium and long term prospects for electric vehicles, and the need for research were reviewed in a comprehensive report by the Select Committee on Science and Technology (House of Lords, 1980). The more recent situation, with particular reference to the "greenhouse effect", has been examined by the Select Committee (House of Lords, 1989 I), and also by the Energy Committee (House of Commons, 1989 I). The Select Committee agrees with the conclusion of the 1980 Report that substitution by electric vehicles, except in limited areas of urban transport, is improbable in the foreseeable future. However, the Electricity Council (1989a) is more optimistic and says that Chloride Silent Power Ltd (a company jointly owned by the Electricity Council and the Chloride Group) is among the leaders in developing the sodium sulphur battery for road transport purposes, and is now constructing a pilot plant as a precursor to full scale production. The new battery will offer the technical capability to operate high performance battery electric vehicles over ranges of at least 100 miles. But the Energy Committee, on the evidence and memoranda submitted by the Electricity Council (1989a and b), think that there are only "very good chances" of the necessary breakthrough in the next 20 years in battery technology which would allow the widespread use of electric vehicles on roads. It seems that there is still no certainty about the key development necessary to give electric vehicles the impetus they need.

Finally, when looking into the long term future (mid to late 21st century), there may be developments in fuel cells which will make electric vehicles of almost unlimited range a viable proposition. The fuel cell is a device which takes in a fuel and an oxidant and converts their reaction directly to electricity. The electro-chemistry of the fuel cell has been described by Bacon (1973), and hydrogen/oxygen fuel cells have been used to provide electrical power on board the US manned space missions. One of their attractions is that the efficiency of producing electricity is not limited by the Carnot cycle to around 40%, as with conventional thermal power stations. An electrically powered city car running from a fuel cell fuelled by compressed hydrogen was demonstrated over 20 years ago, and had a range on one filling of gas of over 200 miles (Bacon, 1973). Hydrogen is not a convenient fuel for mobile use, but while fuel cells can be designed to run on hydrocarbon fuel instead of hydrogen, it may be necessary to react the fuel in a steam reformer, with consequent complications and increased weight and bulk. Some of the complexities of using a methanol fuel cell in a compact car were illustrated in a design study by Appleby and Kalhammer (1980). Considerable increases in weight and cost were predicted, though a fuel consumption of about one third of the petrol version was estimated.

Even the hydrogen fuel cell has to overcome formidable cost and development problems before it can be seriously considered for road transport use, and it seems that it will have to wait for the "hydrogen economy" before finding widespread application. But research work is in progress, and brief descriptions of developments, and present thoughts on fuel cells for transport, are given by Huff *et al* (1987), Hirschenhofer (1989) and Romano (1989).

The widespread use of advanced electric vehicles still seems a long way in the future, but ACEC (1978) saw electric propulsion as "inevitable in the long run, when fossil fuel physically runs out or is constrained by economics, serious environmental restrictions, etc". It may well be that the need to reduce CO_2 and other emissions from road transport will provide the thrust to develop the batteries and other components for an electric vehicle future.

7.4 Summary

Battery electric road vehicles showed great promise in the early years of the century, but rapid development of the internal combustion engine, and slow progress with storage batteries, relegated the battery vehicle to specialised use.

In the 1970's and early 80's, oil price rises and greater stress on reduced emissions led to developments of battery urban light vans and buses. The light vans seemed to come nearest to commercial success. In parallel, new battery concepts attracted research and development funding, though none have yet entered production.

The future prospects of battery (or hybrid) road vehicles appear to depend on continuing pressure to improve air quality in cities, and on successful battery development. Reduction of CO_2, the main greenhouse gas, from a move towards electric vehicles depends on the generation of electricity from fuels other than coal and oil.

In the very long term future, say the middle of the next century, scarcity of natural oil and continuing concern about global warming may make electric vehicles the preferred provider of road transport, though the generation of electricity from non-fossil fuel may have to be accepted.

8. Conclusions and future trends

8.1 Pressures for further fuel economy

It must be admitted that, in the late 1980's, there seemed to be little incentive for improving the fuel economy of road vehicles. The price of natural crude oil was declining, not just in real terms, but in money-of-the-day terms as well, and the threat of scarcity from depletion of reserves seemed to be receding. As a result, in the UK, the fuel economy of new cars entering the fleet no longer improved year by year, and there were indications, as larger engined cars were bought, that the new car consumption might be increasing. There was still a considerable difference between estimates (admittedly rough) of the total car fleet consumption, and new cars, so that overall fuel economy could be expected to improve over the next few years, but long term pressure for improvement seemed to be reducing.

Two factors have made a considerable change in perception of the continuing need to improve the fuel economy of road transport. The first is the appreciation, at the end of the 1980's, that emissions of CO_2 from the burning of fossil fuels was contributing to an enhanced greenhouse effect in the atmosphere. Predictions of global warming, climate change, and sea level rises followed, and while the extent of these serious effects is a matter of scientific controversy, there is now a consensus that global warming is likely as a result of increased CO_2 emissions, and that steps are needed to limit them. As CO_2 is an inevitable result of burning fossil fuel, and road transport (in the UK) contributes about 18% of CO_2 emission at present, with increases expected as traffic is forecast to rise, there is new emphasis on using less fuel in road transport.

The second factor is the realisation (again) of the key part played by the Middle East in oil production, and in the location of oil reserves. The 1990 Gulf crisis, and the military actions taken to liberate Kuwait in 1991, have increased the emphasis given in future planning to security of supply for the Western (and Eastern) developed nations.

It is open to doubt whether both these factors are perfectly reflected in the price mechanism for the longer term. And it is long term developments, both in energy supply and in transport technical development, which will determine how well road transport is able to perform in the future.

8.2 Improving fuel economy - technical factors

For private cars, which use about 70% of road transport fuel in the UK at present, there are a few developments which could save quite substantial quantities of fuel, without restricting car use or performance. Changes which might save 10% or more on each vehicle include:

- A move to much greater use of diesel engines, naturally aspirated or turbo-charged, for cars and light vans
- Improvement to the fuel economy of engines while they are not fully warmed up
- Development of advanced automatic gearboxes which could be programmed for economical driving (an "economy gearbox")
- Down-sizing cars to the smallest and lightest which are compatible with the essential passenger (and luggage) accommodation, and occupant protection in a crash.

Greater use of diesel cars, and down-sizing are the two actions that could save most fuel, making use of technology presently proven and available. However, both possibilities have longer term implications. Greater use of diesel engines could cause a shortage of the diesel fraction of crude oil (and a glut of petrol), though diesel car use would have to increase very greatly for this to happen. Perhaps more serious is the increasingly stringent control of NO_x and particulate emissions which may affect the cost and economy of the diesel engine. Oxidation catalysts are available for diesel cars to reduce particulates: catalysts for reducing NO_x may be more of a problem and NO_x is a precursor of the greenhouse gas, ozone, in the lower atmosphere.

The move to smaller cars, for fuel economy improvement, is likely to make it more difficult to provide the same standard of protection to the occupants in the event of a collision. Providing good occupant protection in small cars is a challenge to vehicle designers.

For heavy goods vehicles, attention to reducing aerodynamic drag would save fuel, especially for the heavier vehicles which do much of their mileage on motorways. The fuel saving would, of course, tend to disappear if running speeds increased, and it could be that the use of cruise speed control devices and speed limiters would help drivers to save fuel. The use of gear change indicators for economy could prove more acceptable to professional drivers having to manipulate ten-speed gear-boxes than to the private car driver.

Emission controls for improving air quality can lead to a penalty in fuel economy - or at best to an improvement in economy which is foregone. While reductions in permitted emissions of carbon monoxide and hydrocarbons can lead to improvements in the combustion process, and thus to better fuel use, restrictions on oxides of nitrogen - an indicator of high temperature combustion - usually lead to worse fuel economy. This is particularly so far diesel engines which run with weak mixtures and a lot of excess air (and nitrogen) present and have high (and efficient) combustion temperatures. The balance between emission controls for air quality, and the effects on fuel use and CO_2 emissions, needs to be continually reviewed when tighter emission controls are proposed. With present (immediate future) controls for cars, fuel economy improvements of less than 15% are predicted by the year 2010, and the fuel economy of goods vehicles is expected to remain static, or improve only slightly.

Much greater improvements in fuel economy than those above have been demonstrated to be technically feasible for cars that satisfy forthcoming exhaust emission standards. These improvements will only be realised in practice if customers become willing to purchase more fuel-efficient cars, even if this implies some reduction in size, maximum speed and acceleration from the typical 2.0 litre car of 1990.

8.3 Improving fuel economy - non-technical factors

Giving the driver information on how his driving technique affects fuel consumption, and some measure of formal training in economical practices can, in principle, reduce fuel consumption by 10% or more, with (in most cases) little increase in journey times. But the motivation to economise with fuel must be there, and it is difficult to persuade car drivers, especially, to modify their style of driving. This, incidentally, is why the "economy gearbox", referred to earlier, may be more acceptable, and thus produce improvements in economy.

For commercial vehicle drivers, training, together with company rules and policy, may make fuel saving a more realistic objective. The cash savings to the company of better fuel economy are more easily seen and publicised than petrol price savings to the private motorist. When combined with technical devices like speed controllers, gear change indicators, and improved aerodynamics, the savings in fuel could be as high as 10% to 15% on the heaviest lorries.

In the use made of the road network, there are several ways in which fairly small savings (up to 5%) can be made by the "fine tuning" of area traffic control systems to optimise fuel use, by traffic management schemes (for example, to replace congested junctions by roundabouts), by providing route guidance for drivers, and by enforcing speed limits on non-built-up roads. The reduction of urban traffic congestion has been predicted to have only a comparatively small effect on fuel use, mainly because most fuel is used in the outer and less congested areas of a conurbation. However, means of smoothing traffic flow to avoid (fuel) wasteful start/stop motoring could help to reduce fuel use as well as reduce travel time, always providing that some form of traffic restraint prevents the road system "filling up" again.

The taxation of road vehicles and fuel (the "fiscal framework") has had a considerable influence on the technical development of engines and vehicles, and on their use. It is only necessary to compare Italy, which has historically had a high fuel tax and has a preponderance of small engined (economical) cars, with the United States, which has the reverse, to see how taxation can influence both design and customer choice. In the UK, motor vehicle taxation has a complex history embracing the Road Fund, where vehicle taxation was earmarked for road construction and maintenance only, and the horse-power tax which had a strong influence on engine design. More recently, the balance of revenue from Vehicle Excise Duty and Fuel Duty has been the subject of debate. For goods vehicles, the concept of paying for road wear caused by high axle loads has resulted in the present system of HGV taxation. It may be that in the future continuing moves towards a fiscal system which makes the user pay through a fuel tax for use of the roads will be seen as an acceptable form of taxation, and help reduce fuel use and CO_2 emissions, thus reducing road transport's contribution to global warming.

Finally, the UK is unusual in having a high proportion of its private car fleet subsidised by business. The so-called "company car" became a favoured business "perk" in the early 1970's when Government incomes policy restricted monetary wage increases in an attempt to reduce inflation. In the early days, company cars were very lightly taxed as far as the user was concerned, and the offer of a company car was seen to be a valuable part of the total "remuneration package". In recent years, tax on the value of the car and its private use has increased considerably. There is some evidence from past research that financial support of company cars encourages more non-business motoring, increases average engine size, distorts the second hand car market, and makes company car users less sensitive to petrol price. The continued increase in taxation until it matches the full value of the car to the recipient can be viewed as one way of encouraging fuel economy amongst a substantial group of motorists.

8.4 The potential of alternative fuels

The prospects for the once favoured medium term alternative - liquids from coal - appear less rosy now that more emphasis is given to reduced CO_2 emissions. The process efficiency for coal conversion of 50%-60% suggests comparatively high overall CO_2 emissions. Fuels with a lower CO_2/heat output, like compressed natural gas may be a way forward, but there are difficulties in using them without incurring cost and weight penalties on the vehicle as well as major changes in the fuel distribution system. In the case of methanol (favoured in the United States), the advantage in CO_2 emissions from the vehicle is only about 7% compared with petroleum, though the fuel has some attractions for efficient engine design which may allow a high fuel economy to be achieved. The use of hydrogen, which produces no CO_2 at all, has superficial attractions, but it is likely to become practical only in the very long term (mid to end of the 21st century). But the hydrogen must be produced from electricity generated without burning fossil fuels, or this stage of the process will contribute to CO_2 emissions. This probably implies a breakthrough in photo-electric cells, or the acceptance of large scale nuclear power.

The battery electric road vehicle has been proposed as a way of continuing to provide transport when oil reserves are diminished, and also of eliminating CO_2 emissions. In the short term, they have to be competitive with today's petrol or diesel powered vehicles, and present state-of-the-art battery electric vehicles are still some way from commercial viability. They may, however, find some limited applications, especially in urban areas, and would benefit in competitiveness from a fairly small improvement in battery performance. It does not seem as though they are likely to reduce CO_2 emissions in total when compared with their diesel powered equivalent, though they can equal the diesel emissions if their electricity is generated from the right mix of primary fuel (eg 100% natural gas; or 50% nuclear/renewable resources and 50% fossil fuel). In the longer term, if batteries now in the R&D stage come to production, more attractive vehicles will become possible, and in the very long term, when natural oil is scarce and expensive, battery vehicles or some hybrid form may become attractive. But if electricity is still generated largely from coal, there may remain problems in containing or reducing emissions of CO_2.

8.5 Fuel economy of other modes of transport

This review has been concerned with the improvement of road vehicle fuel economy, and has concentrated in the main on the private car and the commercial goods vehicle. However, it is frequently claimed that other modes of transport are inherently more energy efficient, and that an "integrated transport policy" should be devised and implemented to encourage, for example, the use of trains and buses for passenger travel, and rail freight as an alternative to the HGV. Is there a well-founded case for this kind of proposal?

Consideration of the energy use of other modes of transport is a major subject, requiring a book on its own. What is attempted here is more modest. Appendix B sets out some of the information from various sources on the energy intensiveness (or specific energy consumption, SEC) of different transport modes. SEC is measured by primary energy used per unit of useful transport work done. While the basic data is straightforward enough, difficulties arise when the number of passengers or payload size has to be taken into account, as it does if any valid comparisons are to be made.

The tentative conclusions (from Appendix B) are that the one area where transfer from private car to public transport (bus or train) would almost certainly save fuel is in the central area of major conurbations. The case for the transfer of other journeys (to public transport) needs to be examined, case by case, with the complete "access and egress" sections included, where relevant. These "access and egress" journeys are important, because few people live, say, at Euston Station, and wish to travel to Birmingham, New Street. The choice, and energy used, for the whole journey can depend crucially on the actual trip origins and destinations, and on the fuel used on the connections to the main public transport interchange.

The same point is also true for freight transport, where the main haul specific energy consumptions of road and rail are not notably different, and the energy used in the "access and egress" stages may change the balance of advantage from one mode to the other in particular cases.

It is, unfortunately, easy to pick values for specific energy consumption of different modes of transport in order to support a piece of special pleading. The difficult part is to make sure that the operational features - load factor, out and return; real origin and destination - which have the major effect on the comparison are chosen fairly. In any case, minimising overall energy use can never be the sole objective. It has, in practice, to be traded against other factors such as cost, time, convenience, reliability and flexibility. Comparisons and policies which are restricted to only part of the whole equation will inevitably produce distortions and inefficiencies.

8.6 Future possibilities

There are no single steps which will enable road transport fuel economy to be improved at a stroke. A very few factors have been identified which might each make reductions of around 10% possible for each vehicle. More have been noted where the effect might be 5% improvement or less. Perhaps the main conclusion to be drawn is that, in future, action on a large number of elements, technical, operational and social will be needed to contain the use of fuel (and of CO_2 emissions) in the expectation of continuing traffic growth.

It has been shown that the technical prospects for greatly improved fuel economy, especially for cars, are there: the challenge for the future is to motivate the customers and operators so that the potential improvements are realised in practice. All the methods of information, education, and taxation policy will be needed to achieve in practice what is possible. There are no magic wands available to vehicle designers or policy makers. The nearest thing might be a means of encouraging private and business motorists to buy fuel-efficient small cars, and then not use them very much.

9. Acknowledgements

It is pleasant to be able to acknowledge the work of friends and colleagues whose research over the past years has formed the basis of this review. Their names appear frequently in the Lists of References.

Thanks are also due to Dr P H Bly, Head of Vehicles Group, TRRL, whose comments have always improved the text, and to members of the Department of Transport Headquarters Divisions who have made constructive criticisms. The staff in TRRL Library have been invariably helpful in locating the references.

The following organisations gave permission for extracts from their publications to be reproduced, and the author is grateful for their help:

The British Petroleum Company plc., London.
Butterworth-Heinemann, Oxford.
The Consumers' Association, London.
The Ford Motor Company Limited, Brentwood.
The Journal of the Institution of Highways and Transportation, London.
The Macmillan Press Ltd, London and Basingstoke.
The Oil and Gas Journal, Houston, USA.
Shell UK Ltd, Cheltenham.

Appendix A
Global warming and road transport*

A.1 The greenhouse effect

The earth is warmed mainly by radiation from the sun. The heat it receives is balanced by heat radiating or reflected back to space by the earth and the earth's atmosphere, so that the earth's temperature remains essentially constant. The sun's radiation, emitted by a very hot body, is of short wavelength: the earth's balancing radiation, from a cool body, is of longer wavelength. Certain gases in the earth's atmosphere are largely transparent to the short wave radiation from the sun, but absorb and re-radiate back to the earth much of the longer wavelength energy emitted by the earth. They trap the heat rather like glass in a greenhouse, so raising the equilibrium temperature of the earth. This is called the "greenhouse effect", and the gases have been called the "greenhouse gases".

Life on earth as we know it depends on the beneficial temperature rise caused by these selective properties of the atmosphere. Without them, average temperatures would be 30°C cooler, the planet covered with ice and any life forms that developed would be very different from the familiar ones around us now.

The balance of energy flows caused by solar radiation and re-radiation from the earth is set out in Table A.1, and illustrated in Figure A.1. The unit of measurement is watts per square metre (Wm^{-2}). Outflows of energy are indicated by negative signs in Table A.1, and inflows by positive numbers. To put the magnitude of solar radiation in perspective, it can be noted that the sun's output is very large compared, say, with the energy output from the use of fossil fuels on earth. For example, in the UK in 1989, the average heat output from all forms of energy use was about 1 Wm^{-2}, compared with 340 Wm^{-2} from the sun.

The Table and Figure show:

(a) Of the 340 Wm^{-2} coming from the sun, 103 Wm^{-2} are reflected or scattered back to space from clouds or the earth's surface (especially ice caps) and therefore play no part in heating the earth or atmosphere.

(b) Of the balance of 237 Wm^{-2}, 64 Wm^{-2} are absorbed by the atmosphere, and 173 Wm^{-2} are absorbed by the earth.

---

* This section is based substantially on the summary of the factors involved in global warming which was published at the end of 1989 in the Report of the House of Lords Select Committee on Science and Technology (House of Lords, 1989 I). Information and comment from other sources are separately acknowledged.

TABLE A.1
Mean global heat balance

	Space	*Atmosphere*	*Earth*
Solar radiation	-340		
(i) absorbed		64	173
(ii) reflected and backscattered	103		
Long-wave radiation			
(i) from the earth's surface	20	365	-385
(ii) from the atmosphere	217	-531	314
Conduction/turbulent convection		102	-102
Total	0	0	0

Units = Wm^{-2} (Watts per square metre)
Source: House of Lords, 1989 I, page 11

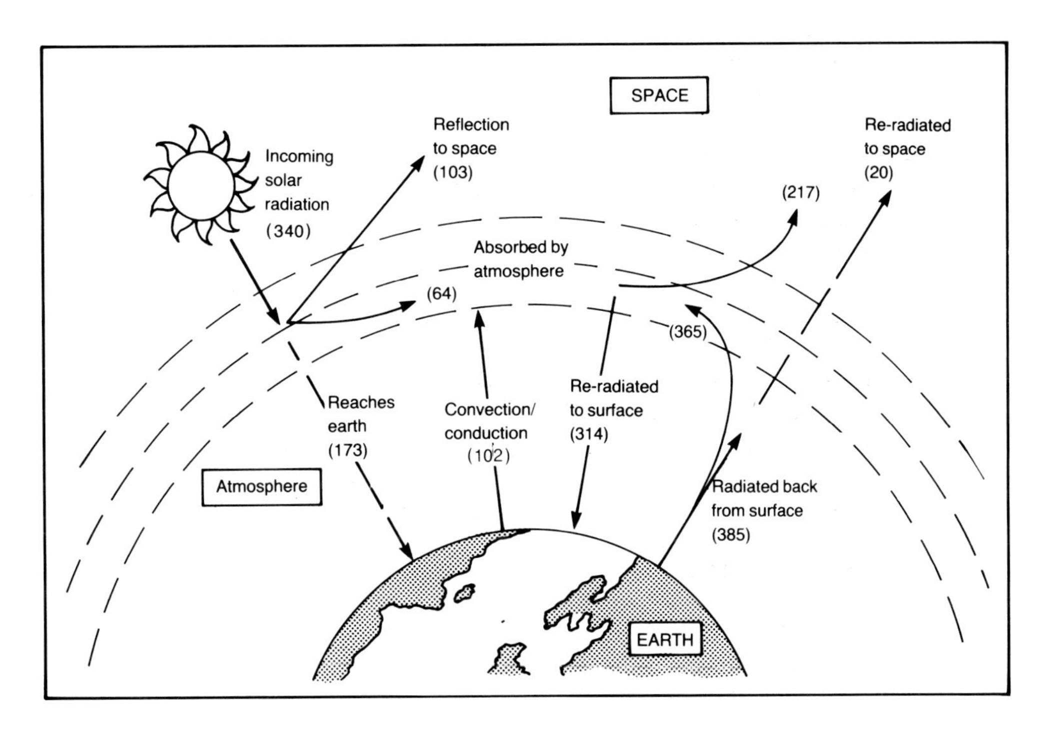

Fig A.1 *Solar and earth radiation balance*
(Source: Energy flows from Table A.1 in Wm^{-2})

(c) Radiation from the earth's surface is at the rate of 385 Wm^{-2}, but only 20 Wm^{-2} are transmitted directly to space through the "atmospheric window". Most (365 Wm^{-2}) is absorbed by naturally occurring gases and clouds in the atmosphere. These gases are principally water vapour and carbon dioxide, with small amounts of methane.

(d) Of 531 Wm^{-2} radiated by clouds and greenhouse gases, 217 Wm^{-2} are radiated upwards to space, and 314 Wm^{-2} are re radiated back to earth.

(e) Contact between the earth and the sea surfaces and the atmosphere results in a transfer from the earth to the atmosphere of 102 Wm^{-2}. This offsets a radiational cooling of the atmosphere of the same amount.

The importance of the natural greenhouse effect is well illustrated by the large proportion (95%) of the long wave radiation emitted by the earth which is absorbed by the atmosphere. The physics of the process is shown in Figure A.2 in terms of radiation from the hot sun and cool earth, and the absorption properties of the main naturally occurring greenhouse gases.

A.2 The greenhouse gases

The main constituents of the atmosphere, oxygen (20%) and nitrogen (80%) play little part in absorbing and re-radiating long wavelength radiation from the earth, as Figure A.2 has shown. The radiatively active gases are water vapour and carbon dioxide and other gases which, by comparison with oxygen and nitrogen, are present in very small quantities. Thus carbon dioxide, even at today's increased concentration, is only present at around 350 parts per million by volume (ppmv) while nitrogen occurs at about 800,000 ppmv. The greenhouse gases are at lower concentrations than the so-called "rare" gases like argon and krypton. The reason why very small quantities of greenhouse gases have such a large effect is because of their absorption of comparatively long wavelength radiation from a cool body like the earth. This can be seen in Figure A.2 where the "atmospheric window" (or Infra Red window) between 7 and 13 mm is very striking.

The radiation from the earth around this waveband, as measured by a satellite, has been compared by Ramanathan (1988) with that which would be expected if there were no absorption by the atmosphere. He shows that there is appreciable atmospheric absorption by water vapour over the range of wavelengths from 6 to 25 mm, by carbon dioxide in the band 12.5-16.7 mm, and to a lesser extent by ozone and methane at 9-10 mm. But nearly 89% of the radiation emitted by the earth in the band between 7 and 13 mm escapes to space. It is because of its transparency to IR radiation that this wavelength band is called the "atmospheric window". Gases like methane, ozone, nitrous oxide, and particularly the synthetic chlorofluorocarbons (CFCs) absorb the earth's radiation strongly in this window region. This has the effect of "dirtying the window", and these gases have a much greater effect, molecule for molecule, than water vapour or carbon dioxide.

Before looking more closely at the magnifying factors of particular gases, it is useful to review the main ones which are usually considered to be the greenhouse gases. Water vapour is not

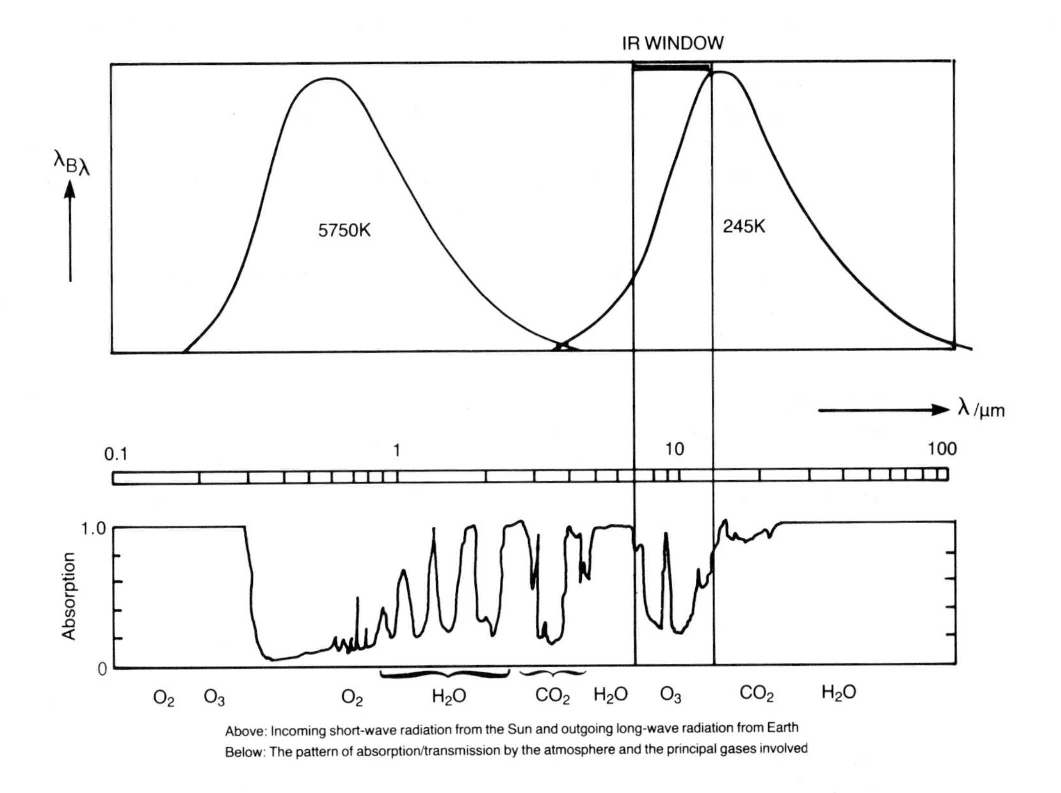

Fig A.2 *Incoming solar radiation and outgoing longwave radiation from the earth (Source: University of East Anglia, 1989)*

always included, but it is by far the most abundant greenhouse gas, and near the earth's surface is present in several thousand parts per million. Its effect is greater than all the other greenhouse gases put together, but the quantity of water on earth means that the concentration of water vapour and the amount of cloud (important for reflecting radiation) are determined by internal mechanisms within the climate system. The other greenhouse gases are largely imposed on the climate system by man's activities, and these are the ones relevant to the present study because changes in activity can change the quantities present in the atmosphere. The gases described below are often referred to for this reason as the "anthropogenic greenhouse gases".

Carbon Dioxide (CO_2). This makes up about 350 ppmv of the atmosphere, and has increased from around 300 ppmv since the beginning of the century. The present rate of increase is about 0.4% per annum. The main source of the increase is the burning of fossil fuels (coal, oil, natural gas), but the reduction of vegetation on the earth's surface, and particularly the destruction of tropical rain forests, have also contributed. Of the carbon released by burning fossil fuels, it is thought that about half remains in the atmosphere and increases the concentration present. The oceans are thought to provide a sink for the remainder, as part of the carbon cycle (Figure A.3) where it can be seen that the flows of carbon are very large compared with the residual quantity added each year to the atmosphere. The mechanism of the cycle is not well understood, and the atmospheric increment is the difference between several large quantities. It may therefore not be

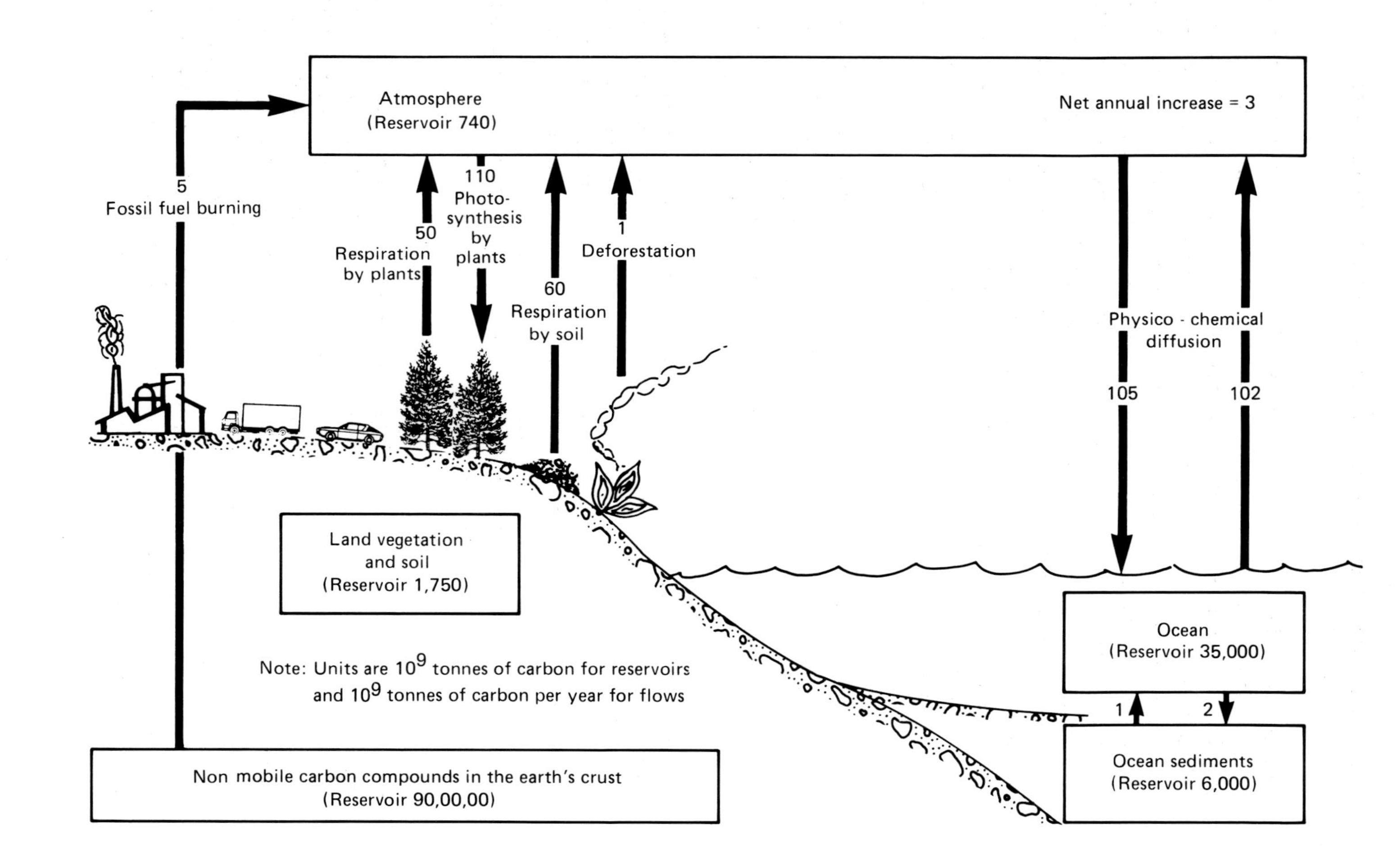

Fig A.3 *The carbon cycle: reservoirs and flows*
(Based on NERC, 1989)

prudent to assume that the proportion of only one half added to the atmosphere will remain unchanged in the future.

Methane (CH_4). This has a concentration of about 1.7 ppmv, increasing at 1.2% per annum. The gas is a natural product of anaerobic decay of vegetation, often associated with marshy areas, and with the digestive processes of grazing animals. It is also released in gas and oil drilling, and leakage during the transmission of natural gas by pipeline. Bolin *et al* (1986) have shown the growth over the past decades from different sources. Over the past century the increase in the gas parallels the growth of the global population, but exploitation of oil and gas reserves has tended to increase the rate of production. Destruction of methane in the atmosphere is believed to be due primarily to instant interaction with the hydroxyl radical (-OH) to form ultimately carbon dioxide and water vapour. While these products are also greenhouse gases, their warming effect is less than that of the original methane because of its strong absorption in the "atmospheric window" region. The destruction of methane can be slowed down if the hydroxyl radicals have been used up in the oxidation of carbon monoxide (CO). Vehicle exhausts are the largest source of CO, and so they may have an indirect effect on methane concentration. (This is discussed later in Section A.4.)

Nitrous oxide (N_2O) This is present at about 0.3 ppmv, increasing at around 0.3% per annum. The gas absorbs strongly in the "atmospheric window" region. It is introduced into the atmosphere by man's activities, mainly through agricultural practices, and also as a result of high temperature combustion - though the most recent information suggests that the importance of combustion as a source has been exaggerated in the past (House of Lords, 1989 II, page 166).

Ozone (O_3). This gas occurs mainly in the upper atmosphere, where it is produced as part of the photo-chemical reaction between the sun's ultra-violet radiation and atmospheric oxygen. It thus helps to protect the earth from dangerous radiation from the sun, and it does not act as a significant greenhouse gas because of its height in the atmosphere. On the other hand, in the lower atmosphere (the troposphere), ozone produced by man's activities does act as a greenhouse gas, with strong absorption in the "atmospheric window" region. Tropospheric concentrations of the gas have been increasing at about 0.25% per annum, to a large extent because of photochemical interactions with oxides of nitrogen (NO_x) and unburnt hydrocarbons in part from motor vehicle exhausts.

Chlorofluorocarbons (CFCs or Freons) are entirely man-made chemicals and are used mainly as refrigerants and propellants in aerosol sprays. They are highly stable compounds, and though only present in the atmosphere in extremely small quantities (about 0.6 parts per billion) have been increasing at around 6% per annum. They have a long lifetime, and are only destroyed in the stratosphere by complex photochemical reactions involving upper atmosphere ozone. There is evidence that CFCs are responsible for the hole in the ozone layer observed over Antarctica. As greenhouse gases they are, volume for volume, 10,000 to 20,000 times more potent than carbon dioxide. However, their potential for damage is so serious that international action (the Montreal Protocol) has been taken to limit their production and use, and there are good reasons to hope that their increasing concentration will be checked.

A.3 Relative importance of the greenhouse gases in global warming

The relative effects of the anthropogenic greenhouse gases on the atmosphere, and hence on global temperature, are estimated by complex radiation calculations. The results are often presented in a simplified way like the "pie" chart in Figure A.4 which gives the relative weighting of the different gases, with carbon dioxide (CO_2) accounting for 50% of the warming effect. The original paper does not put the results in this form, and more significantly makes it clear that the relative effects have changed over different decades, and will change again in the future. This is because the major greenhouse gas (CO_2) is already present in such large quantities that additional increments will tend to give a logarithmic response. Methane and ozone, on the other hand, are present in much smaller quantities at present, and absorb radiation in the region of the "atmospheric window": their influence on temperature is more nearly linear.

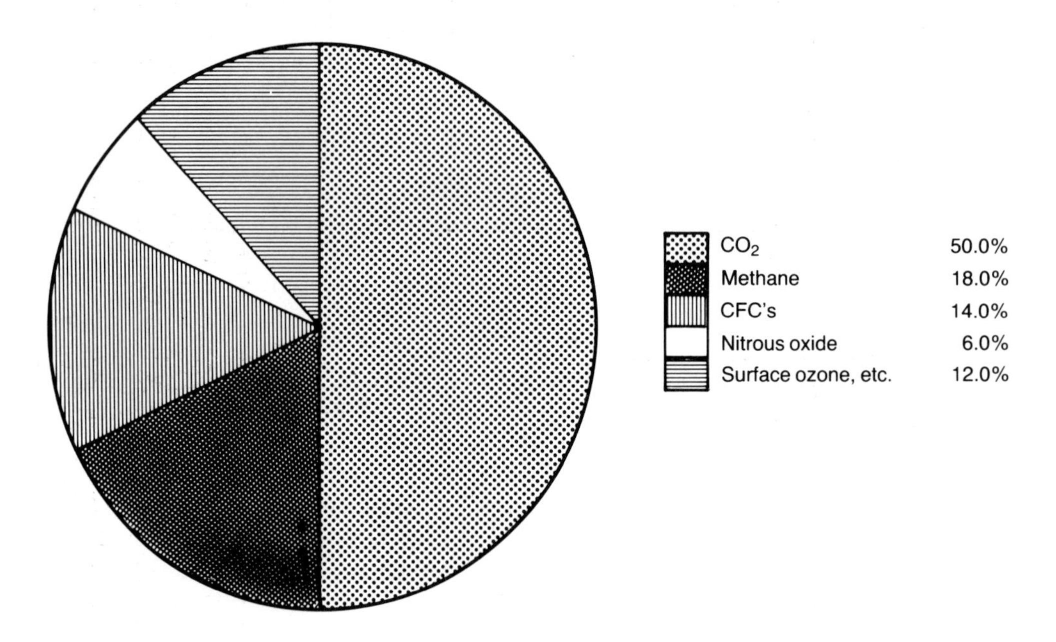

Fig A.4 *Relative contribution of greenhouse gases to global warming 1975-1985 (Adapted from Hansen et al, 1988)*

Hansen *et al* (1988) give the relative effects of the greenhouse gases for various decades in the form of histograms. Figure A.4 is based on the computed effects for the decade 1975-1985. The situation in the past has been assessed, and projections into the future have been made by Warrick *et al* (1990) for the greenhouse gases excluding tropospheric ozone. The results are summarised in Figure A.5. For the period 1950-1985 the estimates agree broadly with Hansen *et al*. The future projections are based on a "business as usual" scenario, with no major action to reduce emissions. In the upper part of Figure A.5, the increasing contribution of the CFCs is evident. The expected results of the banning of CFCs by the Montreal Protocol is seen in the lower chart. Assuming that this is effective, it can be seen that the influence of the CFCs will be reduced (though not eliminated for many years). The atmospheric concentration of CO_2, on these projections, will be about 430 ppmv in the year 2030 compared with 350 ppmv at present.

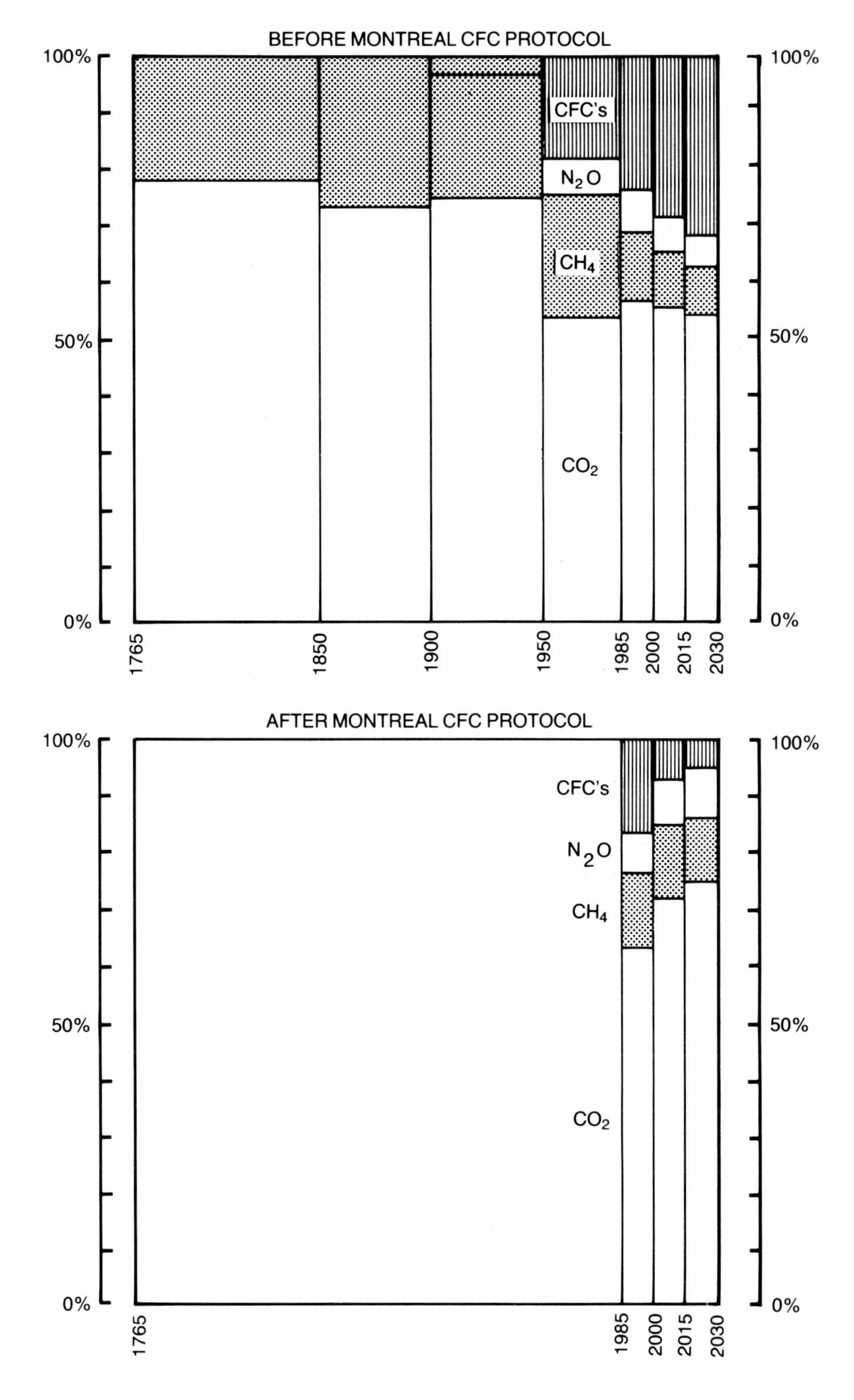

Fig A.5 *Relative contributions to the greenhouse effect 1765-2030 (Source: Warrick et al, 1990)*

TABLE A.2

Percentage of total global temperature rise caused by different greenhouse gases: forecasts from different sources

Period (Source)	Greenhouse Gas CO_2	CH_4	N_2O	O_3	*CFC's*	*Other*
1975-1985						
(1)	49%	19%	5%	8%	17%	2%
Mid 21st Century						
(2)	75%	12%	9%	-	5%	-
(3)	58%	20%	4%	9%	8%	-
(4)	64%	10%	9%	4%	13%	-
(5)	48%	9%	6%	9%	26%	2%
(6)	62%	11%	8%	11%	5%	3%

Sources:

(1) Hansen *et al* (1988)

(2) Warrick *et al* (1990) [O_3 and other gases excluded]

(3) Meteorological Office (1989)

(4) Department of Energy (1989) and Department of the Environment (1989a)

(5) Bolin *et al* (1986) without the reduction of CFC's due to the Montreal Protocol

(6) Bolin *et al* (1986) with adjustment to 5% contribution from CFC's

NOTE: All forecasts, except for (5), assume that the Montreal Protocol has reduced the influence of CFC's

This is the result of one group's look into the future. Other forecasts of the relative importance of the different greenhouse gases have been made, and some of the results are brought together in Table A.2, which includes for completeness Hansen's estimate for the situation during the past decade. Quite large variations in the expected importance of the major gas, CO_2, in the next century are evident, ranging from 75% to 58% (assuming that CFC emissions are controlled by the Montreal Protocol). There is a two to one variation in the expected influence of CH_4 and N_2O. The range for tropospheric O_3 is even larger. The broad conclusions to be drawn are that CO_2 will probably account for around 65% of global warming in the early part of the next century (to 2030), compared with 50% in the past decade: that the effect of CH_4 will be reduced, but that of N_2O will increase somewhat to contribute around 20% in total: and that tropospheric O_3 will contribute about 8% as now. All these forecasts imply a doubling of the effective concentrations of the greenhouse gases by the middle of the next century (Department of the Environment, 1989a).

The complex atmospheric chemistry, and the consequent difficulty of understanding past and present trends so that they can be projected far into the future has been expressed clearly by Thrush (1989) as follows:

> "The relative contributions of different gases to surface warming can be calculated more accurately than the absolute greenhouse effect. In the period from 1850 to about 1960, increased CO_2 was the dominant contributor to any atmospheric warming. In the 1970's and 1980's the combined effects of increased CH_4, N_2O, CFC's, water vapour and O_3 would have been comparable with that of CO_2. The increase in water vapour arises from the increase in temperature. However, water vapour acts both as a greenhouse gas, and chemically by increasing the rates of reactions removing tropospheric CH_4 and tropospheric pollutants such as hydrocarbons. These reactions yield and destroy O_3, another

greenhouse gas, so that their net effect on surface temperature could be negative rather than positive. Present evidence is that these secondary effects are much smaller than the uncertainties in climate modelling."

Another way of looking at the relative influence of the greenhouse gases is to consider them in terms of their equivalent CO_2 concentration. This gives a series of multiplying factors based, again, on concentrations and changes in concentrations usually in the past decade. Table A.3 gives generally accepted values which are based on the work of Ramanthan *et al* (1985). They are broadly consistent with Hansen *et al* (1988) in the overall effect of the different gases, and relate to the period 1975 - 1985. From the previous paragraphs it will be obvious that the multiplying factors are likely to change in the future, and they are not therefore robust for making long term forecasts.

TABLE A.3

Relative greenhouse forcing effects of some trace gases

Gas	*Greenhouse forcing per molecule relative to CO_2*	*Concentration change 1975-85*	*Relative greenhouse forcing 1975-85*
CO_2	1	+ 15 ppm	1.0
CH_4	30	+ 0.17 ppm	0.3
N_2O	160	+ 0.01 ppm	0.1
CFCs	~17,000	+ 0.3 ppb	0.4
Other gases	-	-	0.1

ppm = parts per million of air

ppb = parts per billion of air

Source: CEGB, 1989, based on model calculations by Ramanathan, 1988, and Ramanathan *et al*, 1985

Finally, what is the evidence that global warming is occurring, or is likely to occur, as a result of increases in the quantities of anthropogenic greenhouse gases in the atmosphere? There is some evidence for an increase in global temperature over the past few decades as shown in Figure A.6, and also in Figure A.7 where the observations are compared with model predictions of temperature rise based on two rates of increase in atmospheric CO_2. Measurements of CO_2 in the atmosphere have shown a steady increase since the start of observations in 1950 (Figure A.8). Over a longer time scale, Houghton and Woodwell (1989) compare temperature deviation with estimates of atmospheric concentration of CO_2 and CH_4, and show that there is a tendency for both temperature and gas concentration to increase over the period 1880 to 1980. There is also evidence from the same authors on a time-scale of many thousands of years that the global temperature (as deduced from the proportions of deuterium and an oxygen isotope in ice core samples) has been correlated with atmospheric concentrations of CO_2 and CH_4.

While the shape of the measured temperature rise over the past 100 years does not exactly follow the predicted global warming in Figure A.7, most scientists accept that there is a causal relationship, though necessarily a complex one. Of the organisations giving evidence to the House of Commons and House of Lords Select Committees in 1989, only British Coal (1989) seriously doubted the evidence that global warming was caused by increases in anthropogenic

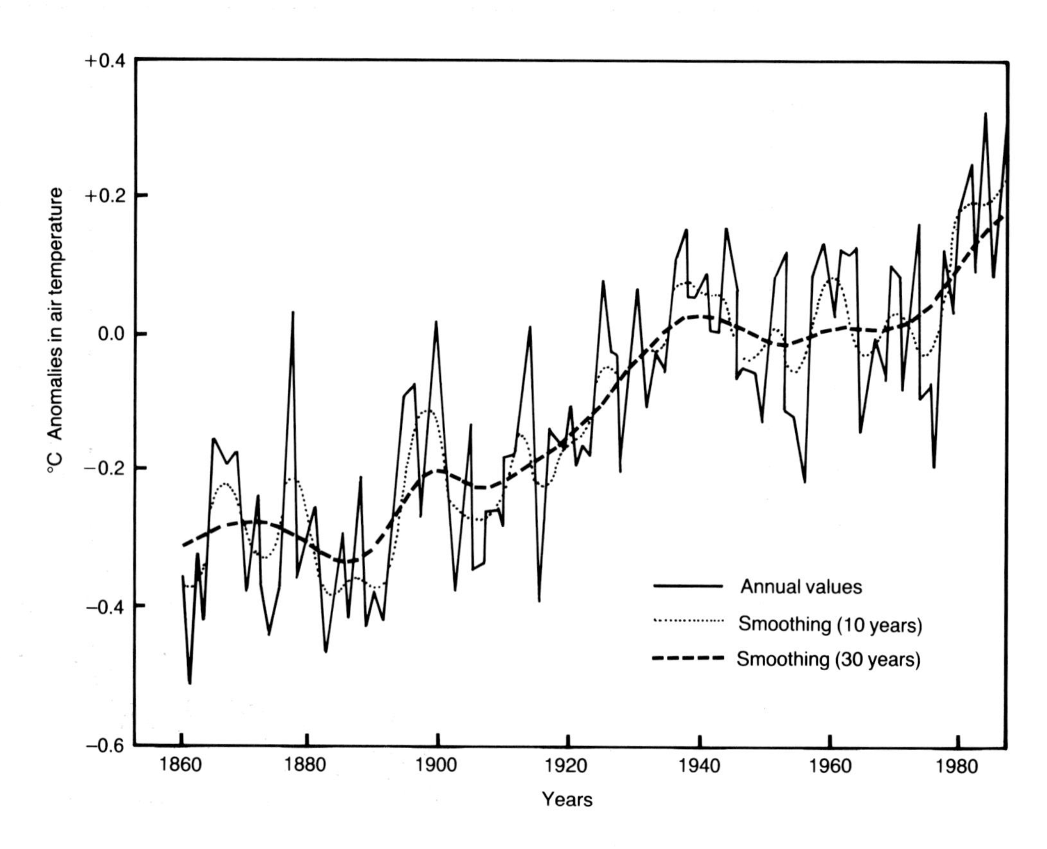

Fig A.6 *Global climatic curves from 1861 to 1987, summarising the air temperature above the land masses and ocean surfaces*
(Source: P D Jones et al. - Nature, Vol. 332, 1988. From House of Commons, 1989 I)

greenhouse gases. Another sceptical source is the George C Marshall Institute, Washington, DC (1989) which points out that the observed global temperature changes could be caused by random climate fluctuations, for example, changes in solar activity which are known to occur on a roughly 200 year cycle. This theory has been the subject of some criticism (Gribben, 1990).

However, the conclusion of the enquiry by the House of Lords (1989 I) is to:

> "... accept the scientific evidence that concentrations of greenhouse gases have increased, and that the global mean temperature is higher now than in 1850. There has been an accelerated warming during the last two decades. (It is considered) ... likely, though not proven, that the observed increase in greenhouse gases has contributed to the warming. But this could be a fortuitous result and due to natural causes."

The Select Committee goes on to say that, based on computer climate models, scientific opinion is that the global mean temperature will increase by 3 ± 1.5°C by some time after 2030. Action,

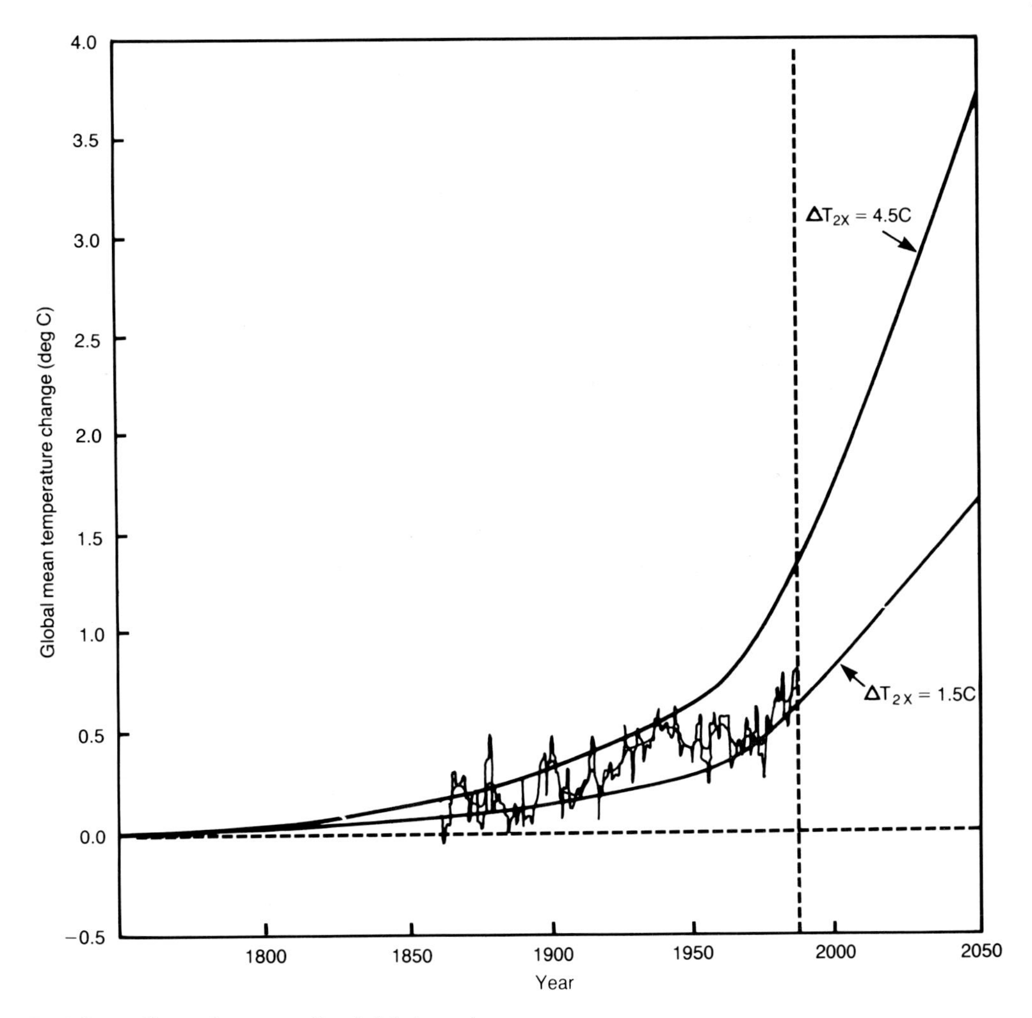

Fig A.7 *Observed versus predicted global warming (Source: University of East Anglia, 1989)*

by way of insurance or "no regrets" policies, is needed in advance of obtaining clear proof that global warming due to enhanced greenhouse gas concentrations is occurring. This is because, by the time clear evidence becomes available, it will be much more difficult to reverse the process.

Work by the Inter-Governmental Panel on Climate Change (IPCC), operating under the auspices of the United Nations Environmental Programme and the World Meteorological Organization, supports this stance. The IPCC Working Group I reached some important conclusions in 1990, including:

- emissions resulting from human activity are substantially increasing the atmospheric concentrations of greenhouse gases.

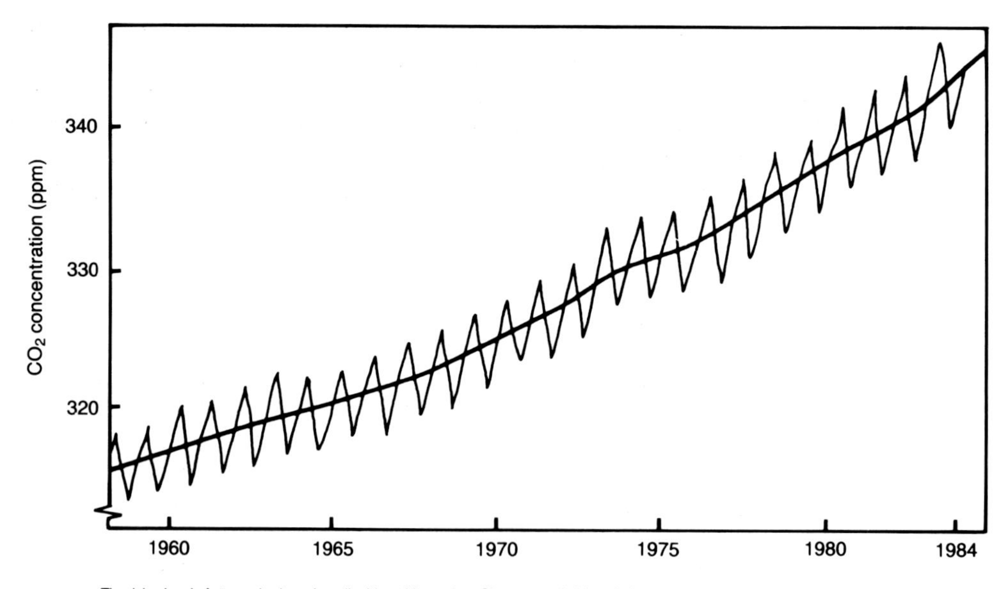

Fig A.8 *Growth of atmospheric carbon dioxide*
(From House of Commons, 1989 I, Source: Bolin, 1986)

- these increases will enhance the greenhouse effect, causing additional warming of the earth's surface.
- carbon dioxide is responsible for over half the man-made greenhouse effect in the past and is likely to remain so in the future.
- continued emissions of many of these gases at present rates would commit us to increased concentrations for centuries ahead.
- the longer emissions continue to increase at present day rates, the greater reductions would have to be made for concentrations to stabilise at a given level.

Using the best available models, Working Group I predicted that, if we continue as we are, global mean temperature would increase by 0.3°C per decade (within a range of 0.2° to 0.5° per decade). While this is less than suggested by the House of Lords Select Committee (above), it is still greater than the world has seen over the past 100,000 years. Rapid changes of climate and sea level would be expected over the next century.

The UK Government accepts that these changes could have major effects on the world. The human consequences, and their effects on international security, could be very severe. The Government believes that this generation has a duty to act to meet this threat of global warming ("This Common Inheritance", 1990).

A.4 Road transport emissions: Past, present, and future

Carbon dioxide has been identified as the main (but not the only) greenhouse gas at present and seems likely to increase in importance into the early decades of the next century. Table A.4 shows the sources of CO_2 emissions in the UK, and the trends since 1977. Road transport accounted for 16% of the total in 1987. (This had increased to 18% in 1988. In the USA, according to Gibbons *et al* (1989), cars and light trucks contributed 15%.) As the UK produced 627 M Tonnes of CO_2 in 1987 - about 3% of the global total emission (House of Lords, 1989 I, page 21) - this means that UK road transport produced about 100 MT of CO_2.

TABLE A.4
Emissions of carbon dioxide from UK sources
(Million Tonnes)

Sector	*1977*	*1980*	*1985*	*1987*	*Per Cent of Total (1987)*
Domestic	85	85	87	87	14
Commercial/public service[2]	34	33	34	32	5
Power stations	238	247	217	233	37
Refineries	25	24	20	21	3
Agriculture (fuel use)	4	3	3	3	1
Other industry[1]	170	136	124	125	20
Rail transport	3	3	2	2	-
Road transport	73	80	88	98	16
Incineration and agricultural burning	6	6	6	6	1
Gas production	1	3	6	8	1
Cement	8	8	7	7	1
Gas flaring	6	8	5	5	1
All sources (ground-based)[2]	654	635	601	627	100

1. Excludes power stations, refineries and agriculture.
2. Excludes emission from fuel used for water and air transport.
Source: Department of the Environment, 1989b

The emissions from road vehicles in the UK of carbon dioxide (CO_2), carbon monoxide (CO), and oxides of nitrogen (NO_x) for the years 1977 - 1987 are shown in Table A.5. Carbon monoxide is not itself a greenhouse gas, but it plays a part in the greenhouse effect by removing hydroxyl radicals (-OH) from the atmosphere, which act as a sink for the greenhouse gas, methane. It has been suggested that the rising level of carbon monoxide has been a factor in the rapid rise in levels of methane by slowing methane breakdown (Greenpeace, 1989). Road transport was responsible for 85% of the carbon monoxide emissions in 1986 (World Wide Fund for Nature, 1989). Oxides of nitrogen (NO and NO_2) are involved in combination with unburnt hydrocarbons and, possibly, carbon monoxide in the formation of tropospheric ozone by the action of sunlight. Road transport contributed 45% of the UK's emissions of NO_x in 1987 (Warren Spring Laboratory, 1989), and 28% of the 2.4 M Tonnes of hydrocarbons in 1987 (Department of the Environment, 1989b). CO, NO_x, and hydrocarbons are all precursors of tropospheric O_3, and therefore add to the greenhouse effect. There is evidence that the concentration of tropospheric ozone doubled in the last 100 years (Meteorological Office, 1989). In some places like Western Europe, California, Eastern USA, and Australia concentrations ten times historic levels have been recorded (Graedel and

TABLE A.5
Emissions from UK road vehicles

	Carbon dioxide Millionsof tonnes			Carbon monoxide Thousands of tonnes			NO_x Thousands of tonnes		
	petrol	*diesel*	*total*	*petrol*	*diesel*	*total*	*petrol*	*diesel*	*Total*
1977	55.69	17.58	73.27	3629	76	3705	432	417	849
1978	58.99	18.32	77.31	3841	78	3919	458	429	887
1979	60.09	18.68	78.77	3911	81	3992	466	442	908
1980	61.56	18.32	79.88	4007	78	4085	477	428	905
1981	60.09	17.22	77.31	3867	75	3942	467	420	887
1982	61.92	17.95	79.87	4187	79	4266	497	435	932
1983	62.66	19.05	81.71	3824	70	3894	462	399	861
1984	64.86	20.88	85.74	3812	72	3884	471	418	889
1985	65.59	21.98	87.57	3934	73	4007	489	426	915
1986	68.89	24.55	93.44	4146	75	4221	520	440	960
1987	71.09	26.38	97.47	4390	80	4470	558	473	1031
% growth 77-87	27%	50%	33%	21%	5%	21%	29%	13%	21%

Source: Greenpeace, 1989, calculated from Parliamentary Answer No 313 30 November 1988

Crutzen, 1989). Oxides of nitrogen, carbon monoxide and man-made hydrocarbons are the chief suspects.

It is difficult to assess how much of UK's contribution to global temperature rise is due to emissions from road transport at present. Road transport CO_2 emissions were 16% of the UK total in 1987, and as CO_2 represents about 50% of the present decade's temperature rise (Figure A.4) road transport CO_2 accounts for about 8% of the total rise. Tropospheric ozone is responsible for about 8% of the overall temperature rise. Road transport emissions, which are precursors of O_3, range from 45% for NO_x to 28% for hydrocarbons. It may be that between one quarter and one half of tropospheric ozone is generated indirectly from road transport emissions. This would add 2% to 4% to the temperature rise, making 10% to 12% of UK's global temperature increase attributable to road transport. But the contribution made by CO, through increasing the concentration of CH_4 is uncertain. Modelling studies imply that between 10% and 40% of the rise in methane may be due to reduction in the natural rate of decay (Department of the Environment, 1989c), and CO from road transport contributes to this rise. The best guess at present is that road transport, in total, accounts for around 10% to 15% of UK global warming, and at least half of this contribution is due to emissions of CO_2.

However, when making projections into the future, it may be acceptable to concentrate on carbon dioxide emissions from road vehicles as their main contribution to global warming. This is because agreement in the EC, and practice in the USA, Japan and other major car producing countries, is making the use of 3-way exhaust catalysts the preferred method of meeting the more rigorous air quality controls from the early 1990s. Thus the emissions of CO, NO_x, and hydrocarbons will be reduced to very low levels, as Table 3.4 has shown. If the 3-way catalyst remains the dominant emission control method it will prevent the realisation of some possible fuel economy gains (and therefore CO_2 reduction) that could otherwise have been achieved by operating at non-stoichiometric air:fuel ratios. This effect has been found by several studies (CCMC, 1987; EC, 1983) to be of the order 5-10% although earlier studies predicted larger penalties. Even so, it may be necessary at some time in the future, to make a choice between further reducing regulated pollutants such as NO_x and CO and containing or reducing emissions of CO_2.

There is also some indication (Matthews, 1990) that CO_2 may have a greater effect than predicted, because of its long life in the atmosphere compared with CH_4, CO, and O_3. Thus carbon dioxide's global warming potential may be larger than previously expected, and it was already forecast to be a larger proportion of the total in the future (65% compared with 50%). If attention is concentrated on CO_2 emissions in the future, the growth of road traffic and the fuel used will be the major factors in determining future global warming.

Traffic forecasts for Great Britain (Department of Transport, 1989) are reproduced in Table A.6, showing projected increases of between 27% and 47% by the year 2000 and between 83% and 142% by the year 2025. No doubt, during this period, there will be changes in vehicle size, fuel efficiency, traffic congestion and other factors that will modify the resulting increase in CO_2 emissions, and the emissions from other, non-traffic, sources will also vary. But substantial reductions in CO_2 emissions per vehicle kilometre would be needed to prevent an increase in the contribution from road transport.

TABLE A.6

Forecasts of road traffic and vehicles in Great Britain

	Actuals			Lower and upper forecasts[1] 1988 = 100			
	1978	1983	1988	1995	2005	2015	2025
(a) Vehicle kilometres							
Cars and taxis	69	78	100	118	140	161	182
				130	168	203	234
Heavy goods vehicles[2]	82	79	100	109	126	144	167
				117	149	189	241
Light goods vehicles	73	75	100	115	139	167	201
				125	170	232	315
Buses and coaches[3]	79	87	100	100	100	100	100
All motor traffic	70	78	100	117	138	160	183
(except two-wheelers)				128	166	203	242
(b) Car[4] ownership							
Cars per person	-	85	100	115	132	147	160
				122	148	168	184
Number of cars	-	84	100	117	137	153	168
				124	153	175	193
(c) Road freight carried by heavy goods vehicles[2]							
Tonne-kilometres	77	74	100	115	139	167	201
				125	170	232	315

1 These figures are taken from *National Road Traffic Forecasts (Great Britain) 1989* (published by HMSO)

2 Over 3.5 tonnes gross vehicle weight

3 Bus and coach traffic is forecast to maintain its 1988 level

4 Body type cars

5 The stock of goods vehicles, buses and coaches is not forecast.

Source: Department of Transport, 1989

Appendix B
Energy considerations for other modes of transport

B.1 The energy intensiveness of transport

The purpose of the Appendix is to present some of the data on energy intensiveness of different modes of ground transport, and to make some comparisons between modes to see what can be learnt about the relative fuel used, and carbon dioxide emissions.

The energy intensiveness (or Specific Energy Consumption (SEC)) of transport is usually measured by primary energy used per useful transport work done. This Appendix summarises the readily available research on the energy intensiveness, or SEC, of the various passenger and freight modes and the factors affecting it. The SEC is commonly measured in units of megajoules per passenger kilometre (MJ/pas.km.) for personal travel. For freight, a common measure is megajoules per payload tonne kilometre (MJ/tonne km). Other units of energy have been used in the past (eg: kWh, litres of fuel, B.Th.U., MTOE, MTCE), and miles used instead of kilometres, but conversion to the preferred units is relatively straightforward.

It is much less straightforward to put the calculated values of energy intensiveness in a form which can allow fair comparison between different modes of transport. An obvious example is car travel. The energy intensiveness of the vehicle on a particular journey depends on the kind of road and type of driver: but it is even more dependent on whether one "passenger" is carried, or four. The load factor for cars differs for different journey purposes, so that any energy intensiveness that is quoted has to have its assumptions carefully defined if it is to be of much use. An early and useful example of the kind of broad comparison that can be made for passenger travel, and the assumptions that are necessary, is reproduced as Table B.1.

A number of estimates of the SEC of different modes are collected together in Table B.2. The figures should be treated as approximate, and for comparative purposes only: the assumptions made are not defined in the Table, and the source references need to be consulted to see what passenger or freight load factors and other conditions are used. With this reservation, the figures do, however, show a degree of agreement, and the most recent ones (Martin and Shock, 1989) are the best, as the original report gives admirable detail on the basis for the values given, as will be seen later.

TABLE B.1

Primary energy consumption by passenger modes and types of traffic

MODE	MJ/seat km[a]	Average[b] load factor	Average MJ/passenger km[a]
PUBLIC TRANSPORT			
Inter-city		%	
Electric loco-hauled train (West coast main line)	0.46-0.47	45	1.0
Diesel loco-hauled train (East coast main line)	0.38-0.41	45	0.9
Express coach	0.20-0.26	65	0.4
Scheduled aircraft	2.2-3.0	65	3.9
Commuter			
Electric multiple unit train (25 kv: 318 seats)	0.30-0.45	25[c]	1.6
Electric multiple unit train (750 v DC: 386 seats)	0.24-0.32	25[c]	1.1
Express bus (50 seats SD)	0.22-0.27	(25)[c][d]	(1.0)[d]
Urban			
Bus (70 seat DD)	0.15-0.23	25[c]	0.8
Underground (LT)	0.20-0.24[e]	14[c][e]	1.6
Rural			
Diesel multiple unit train	0.29-0.36	(20)[d]	(1.6)[d]
Bus (45 seat SD)	0.19-0.23	15	1.4
PRIVATE TRANSPORT			
Car		Average load	
Motorway	0.75-0.80	2 passengers	1.6
Rural	0.80-0.83	(1.7 passengers)[d]	(2.0)[d]
Urban	1.0-1.2	1.5 passengers	3.1
Motorcycle			
Motorcycle (2 seats)	0.94 (average)	(1.1 passengers)[d]	(1.7)[d]
Moped (1 seat)	0.94 (average)	1 passenger	0.9

Notes:

(a) The unit of energy used (MJ) is the megajoule (10^6 joules)
1 Joule = 1 Watt second, 1 kWh = 3.6 MJ = 3412 Btu.

(b) The load factor is the percentage relationship between passengers carried and, normally, seat capacity. The passenger flows in both directions are usually combined. More precisely the load factor, over a defined period of time, is equal to:

$$\frac{\text{Passenger miles}}{\text{seat miles}} \times 100$$

(or sometimes seat miles plus standing capacity miles)

(c) Average load factors for commuter and urban-services conceal large variations between peak and off-peak loadings.

(d) Figures enclosed in brackets are based on assumed load factors.

(e) Energy consumption for London Transport (LT) underground in terms of MJ/place km.

Source: ACEC, 1976

TABLE B.2.
Specific energy consumption for passenger and freight transport (MJ/passenger km)

	Hirst (1973)	ACEC (1976)	Leach *et al.* (1979)	Hillman and Whalley (1983)	Hammar-strom (1988)	Martin and Shock (1989)
PUBLIC PASSENGER TRANSPORT						
Inter-city						
Main-line train	1.1	0.95	-	0.65	1.1	1.2-1.3
Express coach	-	0.40	0.79	-	1.0	0.5-1.0
Commuter						
Rail MU	-	1.35	1.7	1.9	1.3	1.2-1.4
Express bus	-	1.0	0.79	0.3	-	0.5-1.6
Urban bus	-	0.8	-	0.75	0.9	0.3-0.9
Underground	-	1.6	1.7	-	1.3	1.4
Rural						
Rail MU	-	1.6	1.7	-	-	1.2-1.4
Bus	0.7	1.4	0.79	-	-	0.5-1.6
PRIVATE TRANSPORT						
Car						
Motorway	-	1.6	-	-	1.4	-
Rural	2.8	2.0	1.8	-	-	1.3-2.8
Urban	-	3.1	-	3.2	2.2	-
Motor cycle	-	1.7	1.0	1.8	-	-
Moped	-	0.9	-	0.7	-	-
FREIGHT TRANSPORT						
HGVs	-	-	2.35	-	-	0.7-2.0
Service vehicles	-	-	(37.5)	-	-	(3.0-16.4)
Rail	-	-	0.5	-	-	0.6-1.0

Sources: as noted at the heads of columns

B.2 Comparison between modes - passenger travel

The overall Specific Energy Consumption for land travel from 1950 to 1986 is shown in Figure B.1. The total value for all modes is remarkably constant, but it disguises the fall in SEC for rail transport up to 1966 due to the replacement of steam traction by diesel power, and the rise in road passenger transport, probably caused by falling load factors on stage buses. (For private cars, the period from 1970 to 1989 has been shown before in more detail in Figure 2.10.)

For more specific information relating to 1986, the key data have been collected by Martin and Shock (1989), and are presented in Tables B.3 and B.4 for road travel, and in Table B.5 for rail passengers. Car travel, as expected, shows a 2:1 variation in SEC (2.8 to 1.3 MJ/pass.km.) depending on the average number of occupants and the journey purpose (Table B.3). For bus travel, the range is greater (Table B.4): it could be as low as 0.3 MJ/pass.km. for a city centre double decker, or as high as 1.6 MJ/pass.km. for a suburban minibus on a low demand route. But even in this latter case, this is a lower energy intensiveness than the car on an urban shopping trip.

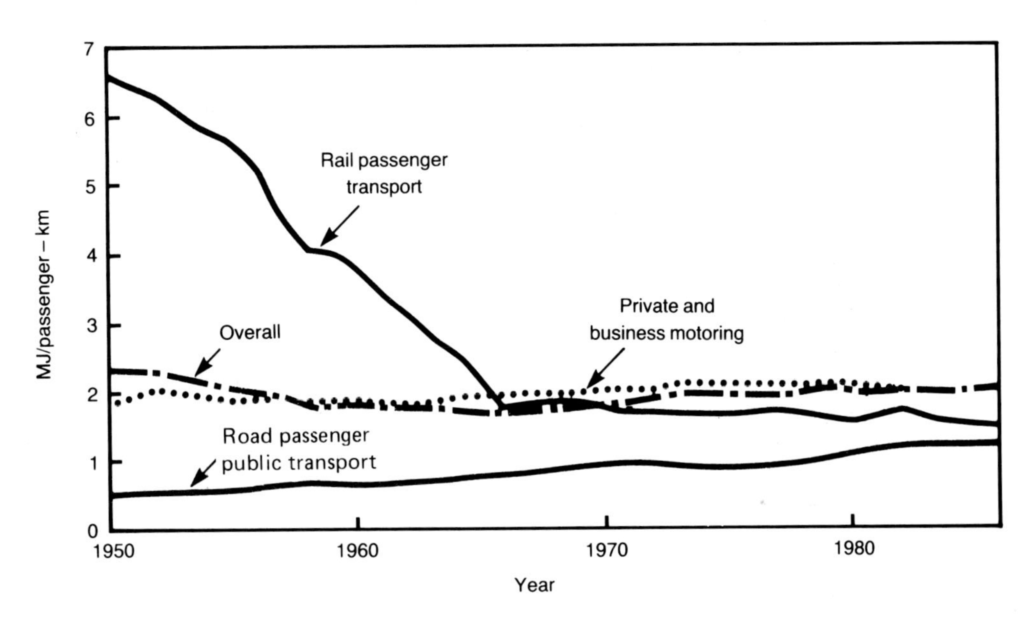

Fig B.1 *Specific Energy Consumption for land passenger transport (1950-1986)*
(Source: Martin and Shock, 1989)

TABLE B.3
Specific energy consumption in 1986 for private and business motoring

Examples of types of journey	*Number of occupants*	*Average SEC (MJ/passenger.km)*
Urban commuting in built-up areas	1.3	2.8
Other urban journeys (eg shopping)	2.0	1.8
Non-urban journeys (eg leisure)	2.3	1.3
Non-urban journeys (eg personal business)	1.7	1.8
Motorway journeys (eg in course of work)	1.1	2.6

Source: Martin and Shock, 1989

Table B.5 gives the SEC figures for rail travel where "average" passenger load factors are very difficult to establish. Even on commuter routes into London at peak times, the average - for both inward and outward directions - is unlikely to be much more than 50%. The figures quoted (1.2 to 1.4 MJ/pass.km.) are not greatly different from bus travel, and are comparable with those achieved by the car in non-urban leisure journeys where car occupancy is high.

B.3 Comparison between modes - freight transport

The variation in energy intensiveness (or SEC) over the past 36 years for surface goods transport is shown in Figure B.2. The same improvement in rail freight SEC as for passenger travel is

TABLE B.4
Specific energy consumption in 1986 for road passenger transport in Great Britain

Vehicle and journey type	*Seating capacity*	*Vehicle occupancy (%)*	*Specific Energy Consumption (MJ/passenger-km)*
Minibus, suburban route	15	25-50	1.6-0.8
Single deck coach, suburban route	33	25-50	1.2-0.6
Single deck coach, motorway route	50	25-50	1.0-0.5
Double deck bus, city centre route	75	50-75	0.4-0.3
Double deck bus, suburban route	75	25-50	0.9-0.5

Source: Martin and Shock, 1989

TABLE B.5
Specific energy consumption in 1986 for rail passenger transport in Great Britain

Fleet type and journey	*MJ/seat-Kilometre*	*Assumed train occupancy (%)*	*Average SEC (MJ/passenger-km)*
Inter-City journey by High Speed Train	0.46	39	1.2
Inter-City journey by electric locomotive	0.53	40	1.3
Provincial/suburban journey by diesel multiple unit	0.30	22	1.4
Provincial/suburban journey by electric multiple unit	0.26	22	1.2
Underground train journey*	0.22	15	1.4

*Applying average figures to seat-km is not very meaningful for the Underground since a high proportion of standing passengers are accommodated at peak periods. Occupancy estimates are shown for place-km including standing passengers rather than seat-km, and the energy use includes the supply of energy for lighting and ventilating the Underground network.
Source: Martin and Shock, 1989

evident up to 1966, again as steam was phased out. The road freight curve shows the increase after 1981 noted previously (Figures 2.11 and 2.12).

The detail for road freight is taken from Martin and Shock (1989), and shown in Table B.6 for heavy goods vehicles. While the variety of vehicle types and duties is extensive (10 separate

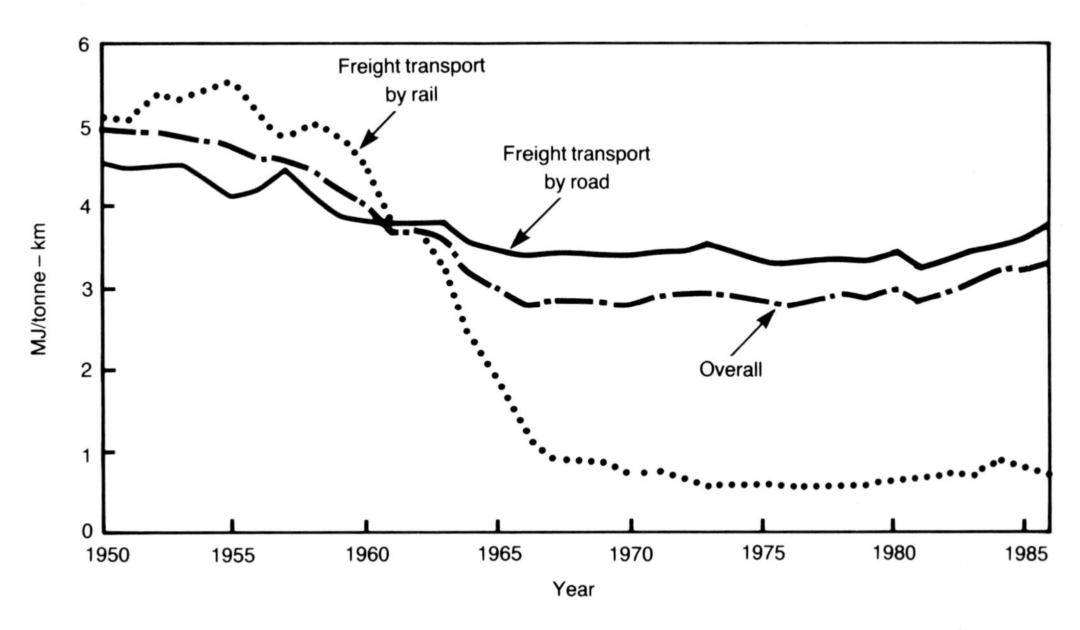

Fig B.2 *Specific Energy Consumption for land goods transport (1950-1986)*
(Source: Martin and Shock, 1989)

headings), the range of SEC is about 3:1 (2.0 to 0.7 MJ/tonne km.), and much less (1.7:1) if two categories concerned with consumer products and component deliveries are excluded. These are likely to be volume limited in load, rather than weight limited as are other commodities, so there is some reason for their high values of SEC.

Table B.7 gives the corresponding data for local delivery and service vehicles. The SEC values are high (up to 7.1 MJ/veh.km.), though other sources have given even higher figures (Table B.2). The lowest SEC for the service vehicles is higher than the highest HGV value. This emphasises the limited use of SEC alone when dealing with a service function, rather than a movement of freight, as Martin and Shock recognised.

The values of SEC for rail freight for various illustrative operations are given in Table B.8. Rather surprisingly, they are not very different from the road freight figures given earlier for HGV's.

B.4 Some comments about modal transfer

The values of energy intensiveness for the different modes of transport and differing operations do not lead at once to an obvious way of saving primary energy, and hence reducing fuel consumption and CO_2 emissions. Even the private car, much reviled, can be competitive with rail on long journeys with car occupancy high. Only in congested urban areas is it clear that public passenger transport has advantages in fuel efficiency over the commuter in his car. Research has

TABLE B.6

Freight transport by road: specific energy consumption for illustrative activities

Illustrative Activity Description	Typical Vehicle Characteristics			Average length of haul km	Typical fuel consumption* litres/100km	Typical SEC MJ/tonne-km (with full load)
	Type	*Weight (permitted gross weight) tonne*	*Payload tonne*			
Long distance movement of building aggregates, minerals etc.	5-axle articulated	38	24	150	43.5	0.7
Haulage of containers	4 axle articulated	32	20	120	35.3	0.7
Movement of export/import goods to/from ports/depots	3-axle articulated	20	12	50	30.0	1.0
Movement of finished goods to distributor	3-axle articulated	20	15	80	29.5	0.8
Movement of semi-finished goods between factories	4-axle rigid	32	18	40	32.0	0.7
Bulk chemical/oil tanker haulage	4-axle articulated	26	18	60	33.0	0.7
Long distance movement of consumer products	4-axle rigid	20	7	300	28.5	1.6
Parcels trunking	3-axle rigid	16	10	40	23.5	0.9
Waste tipping	3-axle rigid tipper	16	10	30	31.0	1.2
Components delivery to assembly works	2-axle rigid	11	4	40	20.5	2.0

*Allowing for part load and empty running

Source: Martin and Shock, 1989

TABLE B.7
Service vehicle transport by road: specific energy consumption for illustrative activities

Illustrative Activity Description	Type	Typical Vehicle Characteristics		Typical fuel consumption litres/100 km	Fuel type	Typical SEC MJ/vehicle km
		Gross weight tonne	*Carrying Capacity tonne*			
High street delivery of goods from local depot	Heavy box van	3.5	1.75	18.5	Diesel	7.1
Utility service work on public highway	Medium	3.0	1.5	12.5	Petrol	4.4
Direct delivery to customers' homes	Medium van	1.75	0.75	15.5	Diesel	6.0
Municipal Council duties	Medium	1.4	0.5	18.5	Petrol	6.4
Utility service work with customer calls	car derived van	1.0	0.25	6.5 10.5	Diesel Petrol	2.5 3.7
Vehicle breakdown repair service	car derived van	1.0	0.25	8.0 11.8	Diesel Petrol	3.1 4.1

Source: Martin and Shock, 1989

TABLE B.8
Rail freight transport: specific energy consumption for illustrative operations

Type of operation	*Average Specific Energy Consumption MJ/tonne-km*
Bulk freight traffic[1] (eg coal delivery by merry-go-round)	0.6
Wagon load traffic	
- single consignments[2]	0.7
- mixed traffic[3]	1.0
Freightliner trunk hauls[4]	0.9

Notes:

1 Merry-go-round trains, handling bulk freight on a continuous basis with special unloading and loading facilities.

2 Trainload trains with large consignments in single trains with little or no intermediate marshalling.

3 Wagonload trains, conveying wagons which require collection from and/or delivery to freight terminals or private sidings, as well as marshalling once or more at intermediate points.

4 Trains carrying containers on trunk hauls between specialised depots.

Source: Martin and Shock, 1989

suggested some strategies which may lead in directions where CO_2 emissions could be contained or reduced (Maltby *et al*, 1978; Leach *et al*, 1979; Hillman and Whalley, 1983):

(a) Policies which encourage free movement of public transport (bus or train) in conurbations, if necessary at the expense of private car use on congested roads.

(b) Assessment of policies for locating commercial and business centres along an existing rail corridor, so that rail load factors can be increased, and car commuting reduced.

(c) Review of planning policies which lead to facilities which are difficult to serve by public transport. Examples include out of town hypermarkets, large district hospitals, and edge-of-town comprehensive schools.

None of these suggestions are technical or operational matters, and, while some of the research is over ten years old, it still appears to be relevant. But the effectiveness of land-use planning to reduce energy use should not be over-estimated. In a comprehensive international study of land-use/transport interactions, the following conclusions about energy savings were drawn (ISGLUTI, 1988):

> "With a given mix of transport technologies, appreciable energy savings can be achieved only by introducing rather oppressive measures which cause travel costs to increase, speeds to fall or car ownership levels to fall, and they may not work if drivers can easily dodge the penalty and continue to use a high-energy mode. Land-use changes seem almost bound to fail in this respect because of the adaptive behaviour of the travellers. This does not mean to say, of course, that putting homes and jobs closer together or speeding up travel is not a worthwhile policy. Those people who rate travel time or cost highly will take advantage of the new situation to reduce their time or money expenditure. Other people will respond by travelling as far as ever, or further if speeds are increased, in order to increase their choice of destination. In total, the reduction in travel time, cost or energy may be very small, but since people have reacted to that improvement in the way that suits them best they have presumably gained the maximum benefit from it (even though their actions may well have diminished the benefits to others, so that the total net benefit is less than could be produced by some hypothetical imposed travel pattern). Even if the most apparent response at present is for people to travel further to gain more choice when an improvement is made, this may not always be the case. Land-use policies may have an important role in keeping options open for the future: if, for example, the costs of travel were to increase greatly in the future (say through higher energy prices) then travellers would be more inclined to sacrifice some of their choice of destination and avail themselves of the potential savings in resource costs which they had previously rejected in favour of more choice."

In order to assess the possibilities again, and as part of a general examination of reducing the demand for travel, the Departments of the Environment and Transport were, in 1990, proposing a joint study of the relationship between land-use development and travel patterns and ways of locating development to reduce travel distances and to increase transport choice ("This Common Inheritance", 1990).

Some of the net savings in energy from assumed changes in travel patterns are shown, for four alternative future scenarios in Figure B.3. Note that these savings are due to the changes in travel which are postulated: the development of policies to bring them about is a different matter, and, as the ISGLUTI conclusions showed, could have all sorts of other effects.

With this proviso in mind, it can be seen in Figure B.3 that significant savings are possible, at least in theory, by "planning" measure like the substitution of business travel by telecommunication, but the highest saving is brought about by a 33% increase in private car fuel economy (mpg). Next is the reduction (by 50%) of discretionary personal travel, and then the transfer of 50% of urban work trips from cars to buses. Maltby's estimates may well need updating, but they are consistent with the modal energy intensiveness values that have been discussed before, and emphasise the importance of direct fuel-saving measures.

One factor which does not appear to have been included in either the earlier or more recent studies, is the effect on fuel consumption of the whole journey from origin to destination, and not just that part on the main mode (eg train). Few passengers live at Euston Station and want to go to Birmingham, New Street. Their actual trip origins and destinations may require relatively inefficient "access and egress" journeys to the main mode of travel. As this part of the whole trip may be in the most congested conditions, high fuel consumption (and CO_2 emissions) may vitiate some of the apparent advantage that the SEC value would have indicated. The same is true for freight transport. It is not known whether studies including these effects have been carried out for energy intensiveness purposes. Access and egress sectors are, of course, a feature of modal choice modelling, both for inter-city travel (WPICT, 1970) and for urban travel (Langdon, 1982), but energy consumption is not included directly - time and cost are the usual main determinants of choice. Studies with the effects on energy (and CO_2 emissions) included could prove a useful contribution to the discussion on the choice of transport mode to improve overall fuel economy.

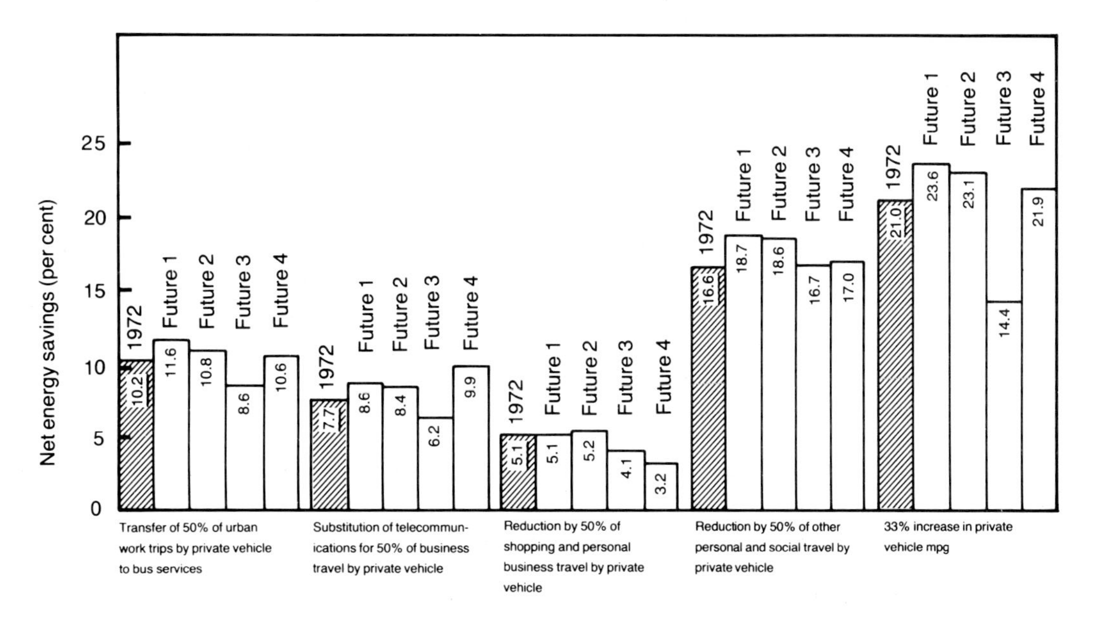

Fig B.3 *Net energy savings as a percentage of transport energy use, if various policy objectives are achieved (Source: Maltby et al, 1978)*

Appendix C
The UK Shell Mileage Marathon

C.1 History of the event

(The following account of the development of the Mileage Marathons is included by kind permission of Shell UK Ltd.)

The history of Mileage Marathons at Shell stretches back over 50 years. They originated at their research laboratory in Illinois in America just before the Second World War with friendly wagers between fellow scientists to see who could get more miles per gallon. From these beginnings with ordinary road cars, with the winner scarcely achieving 50 mpg, more organised competitions evolved. The cars developed too as regards tuning and special preparation so that nearly 150 mpg was obtained after 10 years. In the 1950s and 60s they became highly modified under their near standard external body shapes, and these purpose built vehicles pushed performance up to about 375 mpg. Interchange of staff between laboratories sowed the seed at Shell's Thornton Research Centre in Cheshire and in 1969 there was a competition for ordinary family cars with very limited tuning. This competition still continues but in 1973 there was a break-away movement by some real enthusiasts with a "Special" class - machines built for the sole purpose of obtaining maximum mpg.

Early "Specials" were adapted cycles and motor cycles, with a side-car platform for the drivers, using 50cc Honda 4-stroke moped engines. 450 mpg was the winning consumption in 1973 but, with purpose built chassis coming into use, the 1000 mpg barrier was broken in 1976.

In 1977 Shell UK organised the first competition outside the company, at the Mallory Park racing circuit, essentially for student teams. The Cranfield Institute of Technology won with 1097 mpg. In 1978 the competition grew further and an open all-comers class was introduced. A new world record was set by the overall winners, Ricardo of engine research and development fame, who achieved 1643 mpg.

In 1979, competitors had to complete a 10-mile course at an average speed of at least 15 mph (before then the required speed was 10 mph). The winners, King's College, London, achieved 1684 mpg. The event moved to Silverstone (the motor racing circuit) in 1980 using a 1 mile course back and forth along the pits straight, and the tight curves at each end were blamed by some competitors for King's 1979 record not being broken. Cranfield had a second win with 1465 mpg.

The 1981 event was run over six laps of the full Club circuit without any tight curves, but high winds put paid to any good results being recorded. Talbot were the best student team and Thornhill School the best junior team. Another new feature in 1981 was the involvement of

MIRA (the Motor Industry Research Association) sponsoring an award for aerodynamic design. The winner, Talbot, was chosen after tests had been carried out in the wind-tunnel of the MIRA laboratories, with a wind resistance of 1.6 lbs force at 30 mph. Talbot went on to become outright winners in 1982 with a new UK record of 1926 mpg.

The 1983 event was memorable for the 2000 mpg barrier being well and truly broken by no less than four machines. Winners and new record holders were the schoolboys from Thornhill with a result of 2575 mpg and the pleasure of having triumphed over the technical might of Ford, Shell Research and King's College in the process.

The world record returned to Silverstone in 1984 when Ford Motor Company achieved 3803 mpg, with 2000 mpg becoming fairly commonplace. The 4000 mpg barrier fell in 1985 with 11 year old Daniel Billington piloting the King's College London entry to 4010 mpg.

The next great advance in performance came in 1988 when two Honda technicians from Japan formed Team 1200 and startled the world with a new record of 6409 mpg. They were reticent about how their engine achieved such fuel economy. Team 1200 won in 1989, 1990, and 1991, but it was not until the last year that the 6000 mpg barrier was broken again. Weather conditions, especially high winds and rain, affect the achieved performance, but the 1991 event was run in good weather, with other teams returning high mpg. (The French team which came second achieved 4841mpg.)

The Team 1200 1991 vehicle is shown in Figure C.1.

Fig C1 *1991 Mileage Marathon Winner (6142 mpg)*
(Source: Shell UK Limited)

C.2 Past Winners

1977 Cranfield	1097 mpg
1978 Ricardo	1643 mpg
1979 King's College	1684 mpg
1980 Cranfield	1465 mpg
1981 Cyclone Hovercraft	1309 mpg
1982 Talbot	1926 mpg
1983 Thornhill School DAF Trucks	2575 mpg
1984 Ford UK	3803 mpg
1985 King's College	4010 mpg
1986 Shell Research	3311 mpg
1987 King's College	3804 mpg
1988 Team 1200	6409 mpg
1989 Team 1200	4652 mpg
1990 Team 1200	4734 mpg
1991 Team 1200	6142 mpg

References

Chapter 1 Introduction

C E C (1988). Energy consumption in the European Community's transport sector (1973-1986). Directorate-General for Transport, Statistical Office of the E. C. Commission of the European Communities. April 1988.

HOUSE OF COMMONS (1989 I). Session 1988-89. Energy Committee Sixth Report. Energy Policy Implications of the Greenhouse Effect. Volume I. Report together with the Proceedings of the Committee. Paper 192-I. London: Her Majesty's Stationery Office.

HOUSE OF COMMONS (1989 II). Session 1988-89. Energy Committee Sixth Report. Energy Policy Implications of the Greenhouse Effect. Volume II. Memoranda of Evidence. Paper 192-II. London: Her Majesty's Stationery Office.

HOUSE OF COMMONS (1989 III). Session 1988-89. Energy Committee Sixth Report. Energy Policy Implications of the Greenhouse Effect. Volume III. Minutes of Evidence. Paper 192-III. London: Her Majesty's Stationery Office.

HOUSE OF LORDS (1989 I). Session 1988-1989, 6th Report, Select Committee on Science and Technology. Greenhouse Effect. Volume I. Report. (HL Paper 88-I) London: Her Majesty's Stationery Office.

HOUSE OF LORDS (1989 II). Session 1988-1989, 6th Report, Select Committee on Science and Technology. Greenhouse Effect. Volume II. Evidence. (HL Paper 88-II) London: Her Majesty's Stationery Office.

MARTIN, D J and R A W SHOCK (1989). Energy use and energy efficiency in UK transport up to the year 2010. Report for the Department of Energy by the Chief Scientist's Group, Energy Technology Support Group, Harwell. London: Her Majesty's Stationery Office.

"THIS COMMON INHERITANCE" (1990). This Common Inheritance. Britain's Environmental Strategy. Presented to Parliament by the Secretaries of State for Environment, Trade and Industry, Health, Education and Science, Scotland, Transport, Energy and Northern Ireland, the Ministries of Agriculture, Fisheries and Food and the Secretaries of State for Employment and Wales. Cm 1200. September 1990. London: Her Majesty's Stationery Office.

WPICT (1970). Comparative assessment of new forms of transport. A Report by the Interdepartmental Working Party on Inter-City Transport (Department of the Environment, Department of Trade and Industry, Ministry of Aviation Supply). Reissued in April 1973 as TRRL Supplementary Reports 1, 2 and 3. Transport and Road Research Laboratory, Crowthorne.

Chapter 2 The use of oil-based fuel: Past, present and future

ACEC (1977). Road vehicle and engine design: short and medium term considerations. Advisory Council on Energy Conservation Paper 5, Energy Paper Number 18. Department of Energy. London: Her Majesty's Stationery Office.

BLAND, B H (1984). Effect of fuel price on fuel use and travel patterns. Department of the Environment Department of Transport TRRL Report LR 1114. Transport and Road Research Laboratory, Crowthorne.

BP (1971). BP statistical review of world energy. The British Petroleum Company Ltd. London.

BP (1989). BP statistical review of world energy. July 1989. The British Petroleum Company plc. London.

BP (1990). BP statistical review of world energy. June 1990. The British Petroleum Company plc. London.

CMEO (1990). Private communication. Chief Mechanical Engineer's Office. Department of Transport, 2, Marsham Street, London, SW1P 3EB.

COCHRANE, S and J FRANCIS (1977). Offshore petruleum resources. A review of UK policy. In: Energy Policy, Vol 5 No 1 page 51.

DEPARTMENT OF ENERGY (1976). Digest of United Kingdom energy statistics, 1976. London: Her Majesty's Stationery Office.

DEPARTMENT OF ENERGY (1981). Digest of United Kingdom energy statistics, 1981. London: Her Majesty's Stationery Office.

DEPARTMENT OF ENERGY (1985). Digest of United Kingdom energy statistics, 1985. London: Her Majesty's Stationery Office.

DEPARTMENT OF ENERGY (1989a). Digest of United Kingdom energy statistics, 1989. London: Her Majesty's Stationery Office.

DEPARTMENT OF ENERGY (1989b). Development of the oil and gas resources of the United Kingdom. London: Her Majesty's Stationery Office.

DEPARTMENT OF ENERGY (1990a). Development of the oil and gas resources of the United Kingdom. London: Her Majesty's Stationery Office.

DEPARTMENT OF ENERGY (1990b). Digest of United Kingdom energy statistics, 1990. London: Her Majesty's Stationery Office.

DEPARTMENT OF TRANSPORT (1976). Transport Statistics Great Britain 1965-1975. London: Her Majesty's Stationery Office.

DEPARTMENT OF TRANSPORT (1984a). Second interim report of the Working Group on Fuel Consumption Targets. Vehicle Engineering Development Unit, 2 Marsham Street, London, SW1P 3EB.

DEPARTMENT OF TRANSPORT (1984b). Transport Statistics Great Britain 1973-1983. London: Her Majesty's Stationery Office.

DEPARTMENT OF TRANSPORT (1987). Transport Statistics Great Britain 1976-1986. London: Her Majesty's Stationery Office.

DEPARTMENT OF TRANSPORT (1989). Transport Statistics Great Britain 1978-1988. London: Her Majesty's Stationery Office.

DEPARTMENT OF TRANSPORT (1990). Transport Statistics Great Britain 1979-1989. London: Her Majesty's Stationery Office.

FOLEY, GERALD (1976). The Energy Question. Pelican Book. Penguin Books Ltd., Harmondsworth.

GOLD, THOMAS (1987). Power from the earth. J M Dent and Sons, London.

INSTITUTE OF PETROLEUM (1989). Oil data sheets (various, and up-dates). The Institute of Petroleum, New Cavendish Street, London.

I E A (1984). Fuel efficiency of passenger cars. An I E A Study. International Energy Agency, Organisation for Economic Co-operation and Development. Paris.

LANGLEY, K F (1987). A ranking of synthetic fuel options for road transport applications in the UK. E T S U Report ETSU-R-33 for the Department of Energy, London.

LINSTER, MYRIAM (1989). Background facts and figures. In: Ministerial Session on Transport and the Environment, Background Reports. European Conference of Ministers of Transport. Paris.

O E C D (1988). Transport and the Environment. Organisation for Economic Co-operation and Development. Paris.

OLDFIELD, R H (1980). Effect of fuel prices on traffic.Department of the Environment Department of Transport TRRL Supplementary Report SR 593. Transport and Road Research Laboratory, Crowthorne.

OPENSHAW, KEITH (1980). Woodfuel - a time for reassessment. In: Energy in the Developing World. Oxford University Press.

RICE, PHILLIP (1982). Trends in G.B. vehicle fuel effeciency, 1970-80. In: Traffic Engineering and Control, pages 224-228, April 1982.

RICE, P and P FRATER (1989). The demand for petrol with explicit new car fuel efficiency effects. A UK study 1977-86. In: Energy Economics, April 1989, page 95.

RRL (1965). Research on Road Traffic (page 78). Road Research Laboratory, Department of Scientific and Industrial Research, London: Her Majesty's Stationery Office.

TANNER, J C (1962). Forecasts of future numbers of vehicles in Great Britain. In: Roads and Road Construction. Vol 40 (477) page 263.

TANNER, J C (1974). Forecasts of vehicles and traffic in Great Britain: 1974 revision. Department of the Environment TRRL Report LR 650. Transport and Road Research Laboratory, Crowthorne.

TANNER, J C (1977). Car ownership trends and forecasts. Department of the Environment Department of Transport TRRL Report LR 799. Transport and Road Research Laboratory, Crowthorne.

TANNER, J C (1981). Methods of forecasting kilometres per car. Department of the Environment Department of Transport TRRL Report LR 968. Transport and Road Research Laboratory, Crowthorne.

TANNER, J C (1983). International comparisons of cars and car usage. Department of the Environment Department of Transport TRRL Report LR 1070. Transport and Road Research Laboratory, Crowthorne.

TRRL (1981). Energy used by transport in the United Kingdom (1980). Department of the Environment Department of Transport TRRL Leaflet LF661 Issue 3, Transport and Road Research Laboratory, Crowthorne.

Chapter 3 Road vehicle fuel economy — technical factors

ACEC (1979). Road vehicle and engine design: short and medium term energy considerations. Advisory Council on Energy Conservation Paper 5 (Second Edition). Energy Paper Number 18. Department of Energy. London: Her Majesty's Stationery Office.

ANDRE, M, A J HICKMAN, T J BARLOW, D HASSEL and R JOUMARD (1989). Measurements of the driving behaviour and the vehicle operations in actual uses - method. Report NNE 8905 of the DRIVE project. INRETS/TRRL/CEDIA/TUV Rheinland.

ARMSTRONG, B D (1983). The influence of cool engines on car fuel consumption. Department of the Environment Department of Transport TRRL Supplementary Report SR 822. Transport and Road Research Laboratory, Crowthorne.

AUTOCAR AND MOTOR (1989a). The car in the 90s: Part Three. Page 64, 20 September 1989. Haymarket Magazines Ltd, Teddington, Middlesex.

AUTOCAR AND MOTOR (1989b). Montego 2.0 DSL Turbo Road Test. Page 34, 25 January 1989. Haymarket Magazines Ltd, Teddington, Middlesex.

AUTOCAR AND MOTOR (1989c). Special K. Page 46, 30 August 1989. Haymarket Magazines Ltd, Teddington, Middlesex.

AUTOCAR AND MOTOR (1990). Brave new world. Page 36, 3 January 1990. Haymarket Magazines Ltd, Teddington, Middlesex.

BADER, C (1981). Development experience with hybrid vehicles. Economic and energy considerations. Paper to EVDG (1981).

BAES Jnr.,C F, S E BEALL, D W LEE and G MARLAND (1980). The collection, disposal and storage of carbon dioxide. In: Interactions of energy and climate. Ed. Bach, Pankrath and Williams. Pages 495 to 519. D Riedel, Holland, 1980.

BLACKMORE, D R and A THOMAS, Editors (1977). Fuel economy of the gasoline engine. The Macmillan Press Ltd., Basingstoke and London.

BAUMANN, H-P (1981). Experience with alternative fuel engines and development of a new transmission system to utilise brake recovery energy. Paper to EVDG (1981).

CMEO (1990). Private communication. Department of Transport, 2, Marsham Street, London, SW1P 3EB.

COLWILL, D, A J HICKMAN and VICTORIA H WATERFIELD (1985). Exhaust emissions from cars in service - changes with amendments to ECE Regulation 15. Department of Transport TRRL Supplementary Report SR 840. Transport and Road Research Laboratory, Crowthorne.

CONSUMERS' ASSOCIATION (1988). "Which?" Magazine Car Buying Guide. June 1988. Consumers' Association Ltd., London.

DEPARTMENT OF TRANSPORT (1990). New car fuel consumption. The official figures. October 1990. Issued by the Department of Transport, 2 Marsham Street, London, SW1P 3EB.

EGGLESTON, H S, N GORISSEN, R JOUMARD, R C RIJKEBOER, Z SAMARAS, and Z-H ZIEROCK (1989). CORINAIR Working Group on emission factors for calculating 1985 emissions from road traffic. Volume 1: Methodology and Emission Factors. Report EUR 12260 EN, Directorate-General, Environment, Nuclear Safety and Civil Protection. Commission of the European Communities. Luxembourg: Office for Official Publications of the European Communities.

ENERGY EFFICIENCY OFFICE (1987). Energy efficiency in road transport. Fuel Efficiency Booklet 20. Department of Energy, Thames House South, Millbank, LONDON SW1P 4QJ

ENERGY EFFICIENCY OFFICE (1988). Fuel saving in trucks through aerodynamic styling. Leaflet on Energy Efficiency Demonstration Scheme Project Profile 335. Energy Technology Support Unit, Harwell.

ENERGY EFFICIENCY OFFICE (1990). Fuel savings in trucks through aerodynamic styling. A demonstration at Exel Logistics. ETSU Report ED/286/335. Based on a report prepared by Leyland DAF Technical Centre for the Energy Technology Support Unit, Harwell Laboratory, acting on behalf of the Energy Efficiency Office, Department of Energy, London.

EVDG (1981). Electric Vehicle Development Group Fourth International Conference. Hybrid, dual mode and tracked systems. Peter Peregrinus Ltd., Stevenage, UK.

EVERALL, P F (1968). The effect of road traffic and traffic conditions on fuel consumption. Ministry of Transport RRL Report LR 226. Road Research Laboratory, Crowthorne.

EVERALL, P F and J NORTHROP (1970). The excess fuel consumed by cars when starting from cold. Ministry of Transport RRL Report LR 315. Road Research Laboratory, Crowthorne.

FELGER, G (1987). Engine management systems - a substantial contribution to emission control. Paper C346/87 to the International Conference on Vehicle Emissions and their Impact on European Air Quality. Proceedings of the Institution of Mechanical Engineers, Mechanical Engineering Publications Limited, London.

FENDICK, M St J and B V WOOLFORD (1987). Conformity of production testing. Paper C351/87 to the International Conference on Vehicle Emissions and their Impact on European Air Quality. Proceedings of the Institution of Mechanical Engineers, Mechanical Engineering Publications Limited, London.

FORD (1981). Ford Energy Report, Volume 1. Chapter 7. Ford Motor Company Limited, Brentwood, Essex. Also reprinted as International Association for Vehicle Design, Special Publication SP 1. Inderscience Enterprises Ltd., Great Britain.

FRANCIS, R J and P N WOOLLACOTT (1981). Prospects for improved fuel economy and fuel flexibility in road vehicles. A Report prepared for the Department of Energy by the Energy Technology Support Unit, Harwell. Energy Paper Number 45. London: Her Majesty's Stationery Office.

GARRY, K P (1990a). Wind tunnel measurements of the drag, surface pressure and spray characteristics of two box-bodied articulated goods vehicles. Department of Transport TRRL Contractor Report CR 203. Transport and Road Research Laboratory, Crowthorne.

GARRY, K P (1990b). Wind tunnel measurements of the drag and spray characteristics of articulated tankers. Department of Transport TRRL Contractor Report CR 204. Transport and Road Research Laboratory, Crowthorne.

GYENES, L (1978a). Fuel utilization of articulated vehicles: method of evaluation and data base. Department of the Environment Department of Transport TRRL Supplementary Report SR 422. Transport and Road Research Laboratory, Crowthorne.

GYENES, L (1978b). Fuel utilization of articulated vehicles: effect of gross vehicle weight. Department of the Environment Department of Transport TRRL Supplementary Report SR 424. Transport and Road Research Laboratory, Crowthorne.

GYENES, L (1980). Fuel utilization of articulated vehicles: effect of power train choice. Department of the Environment Department of Transport TRRL Supplementary Report SR 585. Transport and Road Research Laboratory, Crowthorne.

GYNES, L (1990). Feasibility of saving energy from aerodynamic styling of articulated vehicles. Phase I. Monitoring Report. Unpublished TRRL Paper. Transport and Road Research Laboratory, Crowthorne.

GYENES, L, T WILLIAMS and I C P SIMMONS (1979). Power requirements of articulated vehicles under cornering conditions. Department of the Environment Department of Transport TRRL Supplementary Report SR 484. Transport and Road Research Laboratory, Crowthorne.

HAGIN, F and S MARTINI, in consultation with R ZELINKA (1981). The use of hydrostatic braking and flywheel systems in buses (Hydrobus and Gyrobus), their future applications in hybrid electric vehicles to reduce energy consumption, and to increase range and performance. Paper to EVDG (1981).

HICKMAN, A J (1990). Private communication.

HICKMAN, A J and C G B MITCHELL (1989). Technical and economic implications of regulations on air pollution and noise from road vehicles. Background Paper for the OECD/ECMT Special Ministerial Conference on Transport and the Environment. To be published as Department of Transport TRRL Research Report 262, Transport and Road Research Laboratory, Crowthorne.

HICKMAN, A J and T C PEARCE (1988). In-service emissions from fuel injected, lean burn, and three-way catalyst cars - an interim report. TRRL Working Paper Vehicles and Environment Division WP 88/44. Transport and Road Research Laboratory, Crowthorne. (Unpublished)

HICKMAN, A J and M H L WATERS (1991). Improving automobile fuel economy. Paper submitted to the OECD Conference on "Tomorrow's Clean and Fuel-Efficient Automobile." Berlin, March 1991. To be published in "Traffic Engineering and Control".

HOFBAUER, P and K SATOR (1977). Advanced automotive power systems. Part 2. A diesel for a sub-compact car. SAE Paper 770113. Society of Automotive Engineers Inc., Warrendale, Pennsylvania.

HOUSE OF LORDS (1989 II). Session 1988-1989, 6th Report, Select Committee on Science and Technology. Greenhouse Effect. Volume II. Evidence. (HL Paper 88-II) London: Her Majesty's Stationery Office.

INGRAM, K C (1978). The wind-averaged drag coefficient applied to heavy goods vehicles. Department of the Environment Department of Transport TRRL Supplementary Report SR 392. Transport and Road Research Laboratory, Crowthorne.

JARVIS, R P (1984). Fuel economy with small automatic transmissions. International Conference on Fuel Efficient Power Trains and Vehicles. Proceedings of the Institution of Mechanical Engineers, Mechanical Engineering Publications Limited, London.

KING, C S (1984). A car for the nineties: BL's energy conservation vehicle. Proceedings of the Institution of Mechanical Engineers, Part D, Volume 198 No 3 pages 21 to 31. Mechanical Engineering Publications Limited, London.

KORTE, V and D GRUDEN (1987). Possible spark-ignition engine technologies for European exhaust emission legislations. Paper C334/87. International Conference, Vehicle Emissions and their Impact on European Air Quality. Proceedings of the Institution of Mechanical Engineers, Mechanical Engineering Publications Limited, London.

LATHAM, S and P R TONKIN (1988). A study of the feasibility and possible impact of reduced emission level from diesel engined vehicles. Department of Transport TRRL Research Report 158. Transport and Road Research Laboratory, Crowthorne.

McARRAGHER, J S, D J RICKEARD, P PERFETTI, K P SCHUG, D G SNELGROVE, T J AARNINK and J BRANDT (1989). Trends in motor vehicle emission and fuel consumption regulations - 1989 update. Report No 6/89 CONCAWE, The Hague, July 1989.

MARTIN, D J and R A W SHOCK (1989). Energy use and energy efficiency in UK transport up to the year 2010. Report for the Department of Energy by the Chief Scientist's Group, Energy Technology Support Group, Harwell. London: Her Majesty's Stationery Office.

MELLDE, ROLF W, IVA M MAASING and THOMAS B JOHANSSON (1989). Advanced automobile engines for fuel economy, low emissions, and multifuel capability. Annual Review of Energy, Volume 14, pages 425 to 444. Annual Reviews Inc., Palo Alto, California.

MOTOR (1988). The direct approach. Motor Magazine, week ending August 13, 1988. IPC Specialist and Professional Press Ltd., Sutton, Surrey.

NAKAZAWA, N, Y KONO, E TAKAO and N TAKEDA (1987). Development of a braking energy regeneration system for city buses. SAE Paper 872265. Society of Automotive Engineers Inc., Warrendale, Pennsylvania.

NAYSMITH, A (1982). Aerodynamic drag of commercial vehicles. Department of the Environment Department of Transport TRRL Supplementary Report SR 732. Transport and Road Research Laboratory, Crowthorne.

NEWTON, W (1985). Trends in road goods transport 1973 - 1983. Department of Transport TRRL Research Report 43. Transport and Road Research Laboratory, Crowthorne.

NOWOTTNY, P M and E J HARDMAN (1977). Preliminary report on a computer simulation of car driving. Department of the Environment Department of Transport TRRL Supplementary Report SR 325. Transport and Road Research Laboratory, Crowthorne.

OECD (1982). Automobile fuel consumption in actual traffic conditions. A report prepared by an OECD Road Research Group. Organisation for Economic Co-operation and Development, Paris.

PEARCE, T C and M H L WATERS (1980). Cold start fuel consumption of a diesel and a petrol car. Department of the Environment Department of Transport TRRL Supplementary Report SR 636. Transport and Road Research Laboratory, Crowthorne.

PENOYRE, S (1982). A crashworthiness rating system for cars. Proceedings of the Ninth International Conference on Experimental Safety Vehicles. Kyoto, Japan, November 1982. Published by NHTSA, Washington, DC.

RAMSHAW, J and T WILLIAMS (1981). The rolling resistance of commercial vehicle tyres. Department of the Environment Department of Transport TRRL Supplementary Report SR 701. Transport and Road Research Laboratory, Crowthorne.

REDSELL, M, G G LUCAS, and N J ASHFORD (1988). Comparison of on-road fuel consumption for diesel and petrol cars. Department of Transport TRRL Contractor Report 79. Transport and Road Research Laboratory, Crowthorne.

REED, DONALD (1990). CAFE vs Auto Safety. SAE Automotive Engineering, Volume 98, Number 11. Society of Automotive Engineers Inc., Warrendale, Pennsylvania.

RENOUF, M A (1979). Prediction of the fuel consumption of heavy goods vehicles by computer simulation. Department of the Environment Department of Transport TRRL Supplementary Report SR 453. Transport and Road Research Laboratory, Crowthorne.

RENOUF, M A (1981). Analysis of the fuel consumption of commercial vehicles by computer simulation. Department of the Environment Department of Transport TRRL Laboratory Report LR 973. Transport and Road Research Laboratory, Crowthorne.

RICARDO, SIR HARRY and J G G HEMPSON (1968). The high speed internal-combustion engine. Fifth Edition. Blackie and Son Limited, London and Glasgow.

RICHARDSON, R M (1980). Derivation of economy strategies for a stepped ratio automatic transmission. Paper C355/80. Conference on systems engineering in land transport. Proceedings of the Institution of Mechanical Engineers, Mechanical Engineering Publications Limited, London.

SHELL UK (1990). Mileage Marathon 1990. Official Programme. 4 July 1990. Shell UK Ltd., PO Box 9, Winchcombe, Cheltenham, GL54 5YR.

SIMMONS, I C P (1979). Fuel consumption of commercial vehicles: instrumentation and analysis of results. Department of the Environment Department of Transport TRRL Supplementary Report SR 508. Transport and Road Research Laboratory, Crowthorne.

SLUTSKY, SIMON and ENRICO LEVI (1984). Regenerative braking in diesel-powered urban buses. SAE Paper 841690. Society of Automotive Engineers Inc., Warrendale, Pennsylvania.

STUBBS, P W R (1980). The development of a Perbury traction transmission for motor car applications. ASME Paper No.80-C2/DET-59. American Society of Mechanical Engineers, New York.

TRRL (1980). Fuel consumption of diesel and petrol cars. Department of the Environment Department of Transport TRRL Leaflet LF 785 Issue 2. Transport and Road Research Laboratory, Crowthorne.

US CONGRESS (1980). Automobile fuel economy: EPA's performance. Seventeenth Report by the Committee on Government Operations, together with additional views. Union Calendar No 582. 96th Congress, 2nd Session, House Report No 96-948. U.S. Government Printing Office, Washington, D.C., May, 1980.

VOLKSWAGEN (1990). Technology for the future. Booklet published by V A G Public Relations Department, Milton Keynes, May 1990.

VOLVO (1990). Volvo LCP 2000 Light Component Project. Press Release at the OECD/IEA Informal Expert Panel Meeting on Low Consumption/Low Emission Automobiles, Rome, February 1990.

WARREN SPRING LABORATORY (1989). Memorandum to the House of Lords (1989 II, page 280).

WATERS, M H L (1980). Research on energy conservation for cars and goods vehicles. Department of the Environment Department of Transport TRRL Supplementary Report SR 591. Transport and Road Research Laboratory, Crowthorne.

WATERS, M H L (1982). The fuel consumption of diesel and petrol cars and light vans of similar road performance. Paper C106/82. Conference on diesel engines for passenger cars and light duty vehicles. Proceedings of the Institution of Mechanical Engineers, Mechanical Engineering Publications Limited, London.

WATERS, M H L and I B LAKER (1980). Research on fuel consumption for cars. Department of the Environment Department of Transport TRRL Laboratory Report LR 921. Transport and Road Research Laboratory, Crowthorne.

WATKINS, L H (1991). Air pollution from road vehicles. TRRL State of the Art Review 1. Transport and Road Research Laboratory, Department of Transport. London: Her Majesty's Stationery Office.

WATSON, R L, L GYENES and B D ARMSTRONG (1986). A refuelling infrastructure for an all-electric car fleet. Department of Transport TRRL Research Report 66. Transport and Road Research Laboratory, Crowthorne.

WEEKS, R (1979). Comparison of fuel consumption of TRRL VW Golf cars with results produced elsewhere. TRRL Working Paper Assessment Division 79(6). Transport and Road Research Laboratory, Crowthorne. (Unpublished)

WEEKS, R (1981). Fuel consumption of a diesel and a petrol car. Department of the Environment Department of Transport TRRL Laboratory Report LR 964. Transport and Road Research Laboratory, Crowthorne.

WILLIAMS, T and D JACKLIN (1979). Methods of evaluating the effects of aerodynamics on the fuel consumption of commercial vehicles. Department of the Environment Department of Transport TRRL Supplementary Report SR 481. Transport and Road Research Laboratory, Crowthorne.

WILLIAMS, T, I C P SIMMONS, and D J JACKLIN (1981). Fuel consumption testing of heavy goods vehicles. Department of the Environment Department of Transport TRRL Supplementary Report SR 687. Transport and Road Research Laboratory, Crowthorne.

WILLIAMS, T, J RAMSHAW and I C P SIMMONS (1985). Dynamometer tests of the efficiency of a van transmission system. Department of Transport TRRL Research Report RR 10. Transport and Road Research Laboratory, Crowthorne.

WOOD, R A, B R DOWNING, and T C PEARCE (1981). Energy consumption of an electric, a petrol and a diesel powered light goods vehicle in Central London traffic. Department of the Environment Department of Transport TRRL Laboratory Report LR 1021. Transport and Road Research Laboratory, Crowthorne.

Chapter 4 Driver, traffic and fuel economy

AA (1981). Fuel devices - a false economy warns AA. AA Service News Release. 30 March 1981. The Automobile Association, Basingstoke, Hampshire

AA (1983). Are the Government fuel figures playing the numbers game ? Drive and Trail Magazine, April 1983. Automobile Association, Basingstoke, Hampshire.

AA (1991). Private communication. The Automobile Association, Basingstoke, Hampshire.

AKCELIK, R, C BAYLEY, D P BOWER and D C BRIGGS (1983). A hierarchy of vehicle fuel consumption models. Traffic Engineering and Control, October 1983, pages 491-496. Printerhall Ltd., London.

ARMSTRONG, B D (1977). The need for route guidance. Department of the Environment Department of Transport TRRL Supplementary Report SR 330. Transport and Road Research Laboratory, Crowthorne.

AUSTRALIAN NATIONAL ENERGY CONSERVATION PROGRAM (1983). Saving Diesel in road transport. Advisory Booklet 8. Australian Government Publishing Service, Canberra.

AUTOCAR (1982). Real road MPG. The Misers. Week ending 1st May 1982. Haymarket Magazines Ltd, Teddington, Middlesex.

AUTOCAR AND MOTOR (1991). Tin-pot solution? Week ending 6th February 1991. Haymarket Magazines Ltd, Teddington, Middlesex.

BLACKMORE, D R and A THOMAS, Editors (1977). Fuel economy of the gasoline engine. The Macmillan Press Ltd., Basingstoke and London.

C E C (1988). Energy consumption in the European Community's transport sector (1973-1986). Directorate-General for Transport, Statistical Office of the E. C. Commission of the European Communities. April 1988.

C E C (1990). Proposal for a Council Directive amending Directive 70/220/EEC on the approximation of the laws of the Member States relating to measures to be taken against air pollution from motor vehicles. Official Journal of the European Communities, No C81, Vol 33, 30 March 1990, Brussels.

CHANG, MAN-FENG and ROBERT HERMAN (1980). Driver response to different driving instructions: effect on speed, acceleration and fuel consumption. Traffic Engineering and Control, November 1980, pages 545-550. Printerhall Ltd., London.

CLAFFEY, PAUL (1979). Automobile fuel economy and the driver. Transportation Research Record, Volume 739, pages 21-26. National Academy of Sciences, Washington DC.

CONSUMERS' ASSOCIATION (1978). Petrol: can you use less? "Which?" Magazine, January 1978, pages 19-23. Consumers' Association Ltd., London.

CONSUMERS' ASSOCIATION (1983). Cutting your petrol bill. "Which?" Magazine, August 1983, pages 354-357. Consumers' Association Ltd., London.

DAMONGEOT, M (1989). Training of personnel in the freight and passenger road transportation sector. In: Energy efficiency in land transport. Ed. M Roma, Directorate-General for Energy, Commission of the European Communities. Luxembourg.

DEPARTMENT OF TRANSPORT (1984). Second interim report of the Working Group on Fuel Consumption Targets. Vehicle Engineering Development Unit, 2 Marsham Street, London, SW1P 3EB.

DEPARTMENT OF TRANSPORT (1988). AUTOGUIDE. Pilot stage proposals. A consultation document. Issued by Department of Transport, 2 Marsham Street, London, SW1P 3EB

DEPARTMENT OF TRANSPORT (1990a). New car fuel consumption. The official figures. October 1990. Issued by Department of Transport, 2 Marsham Street, London, SW1P 3EB

DEPARTMENT OF TRANSPORT (1990b). Transport Statistics Great Britain 1979-1989. London: Her Majesty's Stationery Office.

DEPARTMENT OF TRANSPORT (1991). Roads Minister publishes draft speed limiter regulations. Press Notice No 134. 17 May 1991. The Department of Transport, 2 Marsham Street, London, SW1P 3EB

DODD, AUSTIN and COLIN WILLIAMS (1981). Study of the fuel economy performance of passenger cars over different driving cycles. Assessment Division Working Paper 81/2. Transport and Road Research Laboratory, Crowthorne. (Unpublished)

ENERGY EFFICIENCY OFFICE (1987). Energy efficiency in road transport. Fuel Efficiency Booklet 20. Department of Energy, Thames House South, Millbank, London, SW1P 4QJ.

EVANS, LEONARD (1979). Driver behaviour effects on fuel consumption in urban driving. Human Factors, Volume 21(4) pages 389-398. The Human Factors Society Inc., Santa Monica, California.

EVERALL, P F (1968). The effect of road and traffic conditions on fuel consumption. Ministry of Transport Road Research Laboratory Report LR 226. Road Research Laboratory, Crowthorne.

FORD (1982). Ford Energy Report, Volume 3, Chapter 24. Ford Motor Company Limited, Brentwood, Essex. Also reprinted as International Association for Vehicle Design, Special Publication SP 1. Inderscience Enterprises Ltd., Great Britain.

GARDINER, P F, R T BAKER and C F LUCAS (1986). Fuel consumption at roundabouts. Department of Transport TRRL Research Report 52. Transport and Road Research Laboratory, Crowthorne.

GUARDIAN (1982). Fool gauge. The Guardian Newspaper, Monday, May 31 1983, page 13. Guardian Newspapers Ltd., London and Manchester.

GYENES, L (1978). Fuel utilization of articulated vehicles: method of evaluation and data base. Department of the Environment Department of Transport TRRL Supplementary Report SR 422. Transport and Road Research Laboratory, Crowthorne.

GYENES, L (1980). Assessing the effect of traffic congestion on motor vehicle fuel consumption. Department of the Environment Department of Transport TRRL Supplementary Report SR 613. Transport and Road Research Laboratory, Crowthorne.

HELLMAN, KARL H and J DILLARD MURRELL (1982). Why vehicles don't achieve the EPA MPG on the road, and how that shortfall can be accounted for. SAE Paper 820791. Society of Automotive Engineers, Inc., Warrendale, Pennsylvania, USA.

HOGBIN, L E and M G BEVAN (1976). Measurement of particulate lead on the M4 motorway at Harlington, Middlesex. Second Report. Department of the Environment TRRL Laboratory Report LR 716. Transport and Road Research Laboratory, Crowthorne.

HUGHES, M T G, D SHAVE and I B LAKER (1988). A set of driving cycles to assess private car fuel efficiency. Department of Transport TRRL Contractor Report CR 81. Transport and Road Research Laboratory, Crowthorne.

JEFFREY, D J (1981). The potential benefits of route guidance. Department of the Environment Department of Transport TRRL Laboratory Report LR 997. Transport and Road Research Laboratory, Crowthorne.

LAKER, I B (1981). Fuel economy - some effects of driver characteristics and vehicle type. Department of the Environment Department of Transport TRRL Laboratory Report LR 1025. Transport and Road Research Laboratory, Crowthorne.

LANGDON, M G (1984). Factors in road design which affect car fuel consumption. Traffic Engineering and Control, April 1984, Printerhall Ltd., London.

LEAKE, G R (1980). Fuel conservation - is there a case for stricter motorway speed limits ? Traffic Engineering and Control, November 1980, Printerhall Ltd., London.

LISTER, R D and R N KEMP (1954). Some traffic and driving conditions which influence petrol consumption. Bulletin of the Motor Industry Research Association, Volume 2, pages 19-26, Northampton, England.

MARTIN, D J and R A W SHOCK (1989). Energy use and energy efficiency in UK transport up to the year 2010. Report for the Department of Energy by the Chief Scientist's Group, Energy Technology Support Group, Harwell. London: Her Majesty's Stationery Office.

McARRAGHER, J S, D J RICKEARD, P PERFETTI, K P SCHUG, D G SNELGROVE, T J AARNINK and J BRANDT (1989). Trends in motor vehicle emission and fuel consumption regulations - 1989 update. Report No 6/89 CONCAWE, The Hague, July 1989.

McNUTT, BARRY D, R DULLA and R CRAWFORD, H T McADAMS and N MORSE (1982). Comparison of EPA and on-road fuel economy - analysis approaches, trends, and impacts. SAE Paper 820788. Society of Automotive Engineers, Inc., Warrendale, Pennsylvania, USA.

MOTOR (1981). Miserly motors. Week ending 7 February 1981. IPC Specialist and Professional Press Ltd., Sutton, Surrey.

MOTORING EXCHANGE AND MART (1991). Classified advertisement. January 24 1991. Link House Advertising Periodicals Ltd., Poole, Dorset.

MULROY, T M (1989). Fuel savings from computerised traffic control systems. In: Energy efficiency in land transport. Ed. M Roma, Directorate-General for Energy, Commission of the European Communities. Luxembourg.

NAYSMITH, A (1989). Energy conservation for car drivers. In: Energy efficiency in land transport. Ed. M Roma, Directorate-General for Energy, Commission of the European Communities. Luxembourg.

OECD (1988). Route guidance and in-car communication systems. Report prepared by an OECD Scientific Experts Group, February 1988. Organisation for Economic Co-operation and Development, Paris, France.

REDSELL, M, G G LUCAS and N J ASHFORD (1988). Comparison of on-road fuel consumption for petrol and diesel cars. Department of Transport TRRL Contractor Report CR 79. Transport and Road Research Laboratory, Crowthorne.

ROBERTSON, D I, C F LUCAS and R T BAKER (1980). Coordinating traffic signals to reduce fuel consumption. Department of Transport TRRL Laboratory Report LR 934. Transport and Road Research Laboratory, Crowthorne.

ROUMEGOUX JEAN-PIERRE (1983). Influence des limitations de vitesse sur la consommation de carburant des vehicules. Recherche - Transports, No.42, April 1983. Institute de Recherche de Transport, Paris, France. (Also as TRRL Translation 3098)

SCOTT, P P and A J BARTON (1976). The effects on road accident rates of the fuel shortage of November 1973 and subsequent legislation. Department of the Environment Department of Transport TRRL Supplementary Report SR 236. Transport and Road Research Laboratory, Crowthorne.

US CONGRESS (1980). Automobile fuel economy: EPA's performance. Seventeenth Report by the Committee on Government Operations, together with additional views. Union Calendar No 582. 96th Congress, 2nd Session, House Report No 96-948. U.S. Government Printing Office, Washington, D.C., May, 1980.

WATERS, M H L (1977). Some sample calculations made in 1973 of fuel savings by speed reduction. Department of the Environment Department of Transport TRRL Internal Note IN 0213/77. Transport and Road Research Laboratory, Crowthorne. (Unpublished)

WATERS, M H L (1981). Meters for encouraging fuel economy in car driving: a review of research requirements. Department of the Environment Department of Transport TRRL Assessment Division Working Paper (81)4. Transport and Road Research Laboratory, Crowthorne. (Unpublished)

WATERS, M H L and I B LAKER (1980). Research on fuel conservation for cars. Department of the Environment Department of Transport TRRL Laboratory Report LR 921. Transport and Road Research Laboratory, Crowthorne.

WATSON, H C, E E MILKINS and G A MARSHALL (1980). A simplified method for quantifying fuel consumption of vehicles in urban traffic. The Journal of the Society of Automotive Engineers - Australasia. January/February 1980. Pages 6-13.

WATSON, R L (1989). Car fuel consumption: its relationship to official list consumptions. Department of Transport TRRL Research Report RR 155. Transport and Road Research Laboratory, Crowthorne.

WEEKS, R (1981). Fuel consumption of a diesel and a petrol car. Department of the Environment Department of Transport TRRL Laboratory Report LR 964. Transport and Road Research Laboratory, Crowthorne.

WOOD, ROGER (1980). Driving patterns of light goods vehicles in urban traffic. Department of the Environment Department of Transport TRRL Supplementary Report SR 607. Transport and Road Research Laboratory, Crowthorne.

WOOD, ROGER and LAURENCE GRIFFIN (1980). The effect of a change in traffic management on fuel consumption. Department of the Environment Department of Transport TRRL Supplementary Report SR 634. Transport and Road Research Laboratory, Crowthorne.

WOOD, ROGER, B R DOWNING and T C PEARCE (1981). Energy consumption of an electric, a petrol and a diesel powered light goods vehicle in Central London traffic. Department of the Environment Department of Transport TRRL Laboratory Report LR 1021. Transport and Road Research Laboratory, Crowthorne.

YOUNG, J C (1988). The influence of road unevenness on vehicle fuel consumption. Paper presented to the PTRC Transport and Planning Summer Annual Meeting, Bath, England. September 1988.

Chapter 5 Taxation and fuel economy

AUTOCAR AND MOTOR (1989). Special K. Page 46, 30 August 1989. Haymarket Magazines Ltd., Teddington, Middlesex.

BUCHANAN, C D (1958). Mixed blessing. The motor in Britain. Leonard Hill (Books) Ltd., London.

CHARLESWORTH, GEORGE (1987). A history of the Transport and Road Research Laboratory, 1933 - 1983. Avebury, Gower Publishing Co. Ltd., Aldershot, Hampshir, and Vermont, USA.

DANIELS, J R (1970). Accelerating progress. Autocar Magazine, 12 November 1970. Reprinted in Motoring Milestones, an Autocar Special, 1979. I P C Transport Press Ltd., Stamford Street, London.

DEPARTMENT OF TRANSPORT (1990a). Transport Statistics Great Britain 1979-1989. London: Her Majesty's Stationery Office.

DEPARTMENT OF TRANSPORT (1990b). The allocation of road track costs, 1990/91, United Kingdom. Issued by the Department of Transport, 2 Marsham Street, London, SW1P 3EB. April 1990.

DEPARTMENT OF TRANSPORT (1991). Malcolm Rifkind proposes initiatives to increase rail traffic: measures to tackle road and urban congestion. Press Notice No 147. 28 May 1991. Department of Transport, 2 Marsham Street, London, SW1P 3EB.

GENERAL ACCIDENT (1990). Company car drivers set bad example to other road users, says Gallup Poll. Press release, 6 December 1990. General Accident Fire and Life Assurance Corporation Ltd., Perth, Scotland.

HAMER, MICK (1991). Pricing cars off city streets. New Scientist, 2 March 1991, New Science Publications, Holborn Publishing Group, London.

HOPKIN, JEAN M (1986). The transport implications of company-financed motoring. Department of Transport TRRL Research Report RR 61. Transport and Road Research Laboratory, Crowthorne.

HOUSE OF COMMONS (1989 II). Session 1988-89. Energy Committee Sixth Report. Energy Policy Implications of the Greenhouse Effect. Volume II. Memoranda of Evidence. Paper 192-II. London: Her Majesty's Stationery Office.

I E A (1984). Fuel efficiency of passenger cars. An I E A Study. International Energy Agency, Organisation for Economic Co-operation and Development. Paris.

INLAND REVENUE (1990). Taxation of mileage allowances (Fixed Profit Car Scheme). Leaflet FPCS 1, Inland Revenue PAYE.

MOGRIDGE, M J H (1985). The effect of company cars upon the secondhand market. Department of Transport TRRL Contractor Report CR 10. Transport and Road Research Laboratory, Crowthorne.

NEWCOMB, T P and R T SPURR (1989). A technical history of the motor car. Adam Hilger, Bristol and New York.

OPEN UNIVERSITY (1989). Memorandum 20 to House of Commons (1989 II, page 140)

PLOWDEN, WILLIAM (1971). The motor car and politics, 1896 - 1970. The Bodley Head, London.
STARK, R H (1983). Diesel cars. Paper presented to the Fleet Management Conference by the Physical Distribution Director, Scottish and Newcastle Breweries (Services) Ltd., Dorset Place, Edinburgh, Scotland.
TANNER, J C (1983). International comparisons of cars and car usage. Department of the Environment Department of Transport TRRL Report LR 1070. Transport and Road Research Laboratory, Crowthorne.
"THIS COMMON INHERITANCE" (1990). This Common Inheritance. Britain's Environmental Strategy. Presented to Parliament by the Secretaries of State for Environment, Trade and Industry, Health, Education and Science, Scotland, Transport, Energy and Northern Ireland, the Ministries of Agriculture, Fisheries and Food and the Secretaries of State for Employment and Wales. Cm 1200. September 1990. London: Her Majesty's Stationery Office.

Chapter 6 Alternative energy sources for road transport

ACEC (1978). Energy for transport: long-term possibilities. Advisory Council on Energy Conservation, Paper 8. Energy Paper Number 26. Department of Energy. London: Her Majesty's Stationery Office.
ARCHER, Dr MARY (1989). Written evidence to House of Lords (1989 II page 284).
BILLINGS, ROGER E (1976). Hydrogen fuel in the sub-compact automobile. SAE Paper 760572. Society of Automotive Engineers Inc., Warrendale, Pennsylvania.
BRITISH GAS PLC (1989). Memorandum 4 to House of Commons (1989 II page 20).
BP (1989). BP statistical review of world energy. July 1989. The British Petroleum Company plc. London.
CHAPMAN, P F, G LEACH and M SLESSER (1974). The energy costs of fuel. Energy Policy, September 1974. Butterworth Scientific Ltd., London.
CHAPMAN, P, G CHARLESWORTH and M BAKER (1976). Future transport fuels. Department of the Environment TRRL Supplementary Report SR 251, Transport and Road Research Laboratory, Crowthorne.
DAVIES, G O (1985). Transport fuels from coal. In: Highways and Transportation. The Journal of the Institution of Highways and Transportation, December 1985, page 2. Institution of Highways and Transportation, London.
DEPARTMENT OF ENERGY (1988). Renewable energy in the UK. The way forward. Energy Paper 55. Quoted in House of Commons (1989 II, page 36).
DEPARTMENT OF ENERGY (1989a). ETSU and IEA papers referred to in supplementary and additional information to House of Commons (1989 III, page 46).
DEPARTMENT OF ENERGY (1989b). Supplementary and additional information to House of Commons (1989 III, page 49).
FORD (1981). Ford Energy Report, Volume 1, Chapter 2. Ford Motor Company Limited, Brentwood, Essex. Also reprinted as International Association for Vehicle Design, Special Publication SP 1. Inderscience Enterprises Ltd., Great Britain.
GAO (1988). Energy security. An overview of changes in the world oil market. Report to Congress. GAO/RCED-88-170 August 1988. United States General Accounting Office, Washington, DC.

GAVAGHAN, HELEN (1989). Bush drives for clean air. New Scientist 7 October 1989. New Science Publications, Holborn Publishing Group, London.

GRAY, CHARLES L Jnr and JEFFREY A ALSON (1989). The case for methanol. Scientific American, November 1989, pages 86-92. Scientific American Inc., New York.

HAMA, J, Y UCHIYAMA and Y KAWAGUCHI (1988). Hydrogen-powered vehicle with metal hydride storage and D.I.S. engine system. SAE Paper 880036. Society of Automotive Engineers Inc., Warrendale, Pennsylvania.

HOUSE OF COMMONS (1989 I). Session 1988-89. Energy Committee Sixth Report. Energy Policy Implications of the Greenhouse Effect. Volume I. Report together with the Proceedings of the Committee. Paper 192-I. London: Her Majesty's Stationery Office.

HOUSE OF COMMONS (1989 II). Session 1988-89. Energy Committee Sixth Report. Energy Policy Implications of the Greenhouse Effect. Volume II. Memoranda of Evidence. Paper 192-II. London: Her Majesty's Stationery Office.

HOUSE OF COMMONS (1989 III). Session 1988-89. Energy Committee Sixth Report. Energy Policy Implications of the Greenhouse Effect. Volume III. Minutes of Evidence. Paper 192-III. London: Her Majesty's Stationery Office.

HOUSE OF LORDS (1989 I). Session 1988-1989, 6th Report, Select Committee on Science and Technology. Greenhouse Effect. Volume I. Report. (HL Paper 88-I) London: Her Majesty's Stationery Office.

HOUSE OF LORDS (1989 II). Session 1988-1989, 6th Report, Select Committee on Science and Technology. Greenhouse Effect. Volume II. Evidence. (HL Paper 88-II) London: Her Majesty's Stationery Office.

KEMPE'S (1989). Kempe's Engineers Year Book, 94th Edition, Volume 1, 1989. Morgan-Grampian Book Publishing Co Ltd, London.

LANGLEY, K F (1983). The future role of hydrogen in the UK energy economy. ETSU Paper R 15, Department of Energy. London: Her Majesty's Stationery Office.

LANGLEY, K F (1987). A ranking of synthetic fuel options for road transport applications in the UK. E T S U Report ETSU-R-33 for the Department of Energy, London.

MAN (1991). MAN Nutzfahrzeuge AG awarded large order - 307 city buses for Australia. 100 vehicles with environmentally compatible natural gas engines. News Release. 25 April 1991. Marketing Department, MAN, Swindon.

MENRAD, H, W LEE and WINIFRED BERNHARDT (1977). Development of a pure methanol fuel car. SAE Paper 770790. Society of Automotive Engineers Inc., Warrendale, Pennsylvania.

NATIONAL COAL BOARD (1978). Liquids from coal. A National Coal Board Report prepared by the Planning Assessment and Development Branch of the Coal Research Establishment, with the Operational Research Executive. National Coal Board, London.

NEUMANN, K H (1989). Energy efficiency improvement as a tool for influencing future engine technology. In: Energy efficiency in land transport. Ed. M Roma, Directorate-General for Energy, Commission of the European Communities. Luxembourg.

PORTER, J (1979). The transition to road transport fuels from coal: a preliminary study. Department of the Environment Department of Transport TRRL Supplementary Report SR 519. Transport and Road Research Laboratory, Crowthorne.

PORTER, J and J W FITCHIE (1977). Energy for road transport in the United Kingdom. Department of the Environment Department of Transport TRRL Supplementary Report SR 311. Transport and Road Research Laboratory, Crowthorne.

UKAEA (1989). Supplementary information from the United Kingdom Atomic Energy Authority to House of Commons (1989 III, Annex page 99).

WATSON, H C, E E MILKINS, W R B MARTIN and J EDSELL (1984). An Australian hydrogen car. Paper presented to the Fifth World Hydrogen Energy Conference, Toronto, 1984.

WELSBY, J K (1974). Fossil fuel supplies up to the year 2000: a preliminary investigation. Department of the Environment TRRL Supplementary Report SR 33. Transport and Road Research Laboratory, Crowthorne.

Chapter 7 Electric road vehicles: An alternative?

ACEC (1978). Energy for transport: long-term possibilities. Advisory Council on Energy Conservation, Paper 8. Energy Paper Number 26. Department of Energy. London: Her Majesty's Stationery Office.

ADAMS, D S (1979). Electric road vehicles - their short and long term future. Paper 9 in Symposium on energy and road transport held at the Transport and Road Research Laboratory, Crowthorne, on 12-13 April 1978. Department of the Environment Department of Transport TRRL Supplementary Report SR 447. Transport and Road Research Laboratory, Crowthorne.

APPELBY, A J and F R KALHAMMER (1980). The Fuel Cell: a practical power source for automotive propulsion ? Paper presented to the "Drive Electric 80" Conference. Drive Electric 80, 30, Millbank, London, SW1P 4RD.

AUTOCAR AND MOTOR (1990a). Los Angeles to turn electric. Page 12, 27 June 1990. Haymarket Magazines Ltd., London.

AUTOCAR AND MOTOR (1990b). Electric shock. Page 50, 18 April 1990. Haymarket Magazines Ltd., London.

BACON, F T (1973). The Melchett Lecture, 1973: Fuel cells and the growing energy problem. Journal of the Institute of Fuel, September 1974, pages 147-162. Institute of Energy, London.

BARAK, M, Editor (1980). Electrochemical power sources. Primary and secondary batteries. IEE Energy series 1. Peter Peregrinus Ltd., Stevenage, UK, on behalf of the Institution of Electrical Engineers, London and New York.

BRUSAGLINO, G and L MAZZON (1989). Electric power vehicles for public service - the outlook for development. In: Energy efficiency in land transport. Ed. M Roma, Directorate-General for Energy, Commission of the European Communities. Luxembourg.

COST 303 (1987). Technical and economic evaluation of dual-mode trolleybus programmes. Final Report and Summary of the Brussels Seminar on dual-mode trolleybuses. Edited by F Fabre and A Klose, Commission of the European Communities. Luxembourg: Office for Official Publications of the European Communities.

DELL, R M (1984). Advanced traction batteries. Paper to University of Strathclyde (1984).

ELECTRICITY COUNCIL (1980). The Enfield car project. Report issued October 1980. Electricity Council, Millbank, London.

ELECTRICITY COUNCIL (1989a). Memorandum 7 to House of Commons (1989 II, page 40).

ELECTRICITY COUNCIL (1989b). Evidence to House of Commons (1989 III, page 24, Question 108).

EVDG (1981). Electric Vehicle Development Group Fourth International Conference. Hybrid, dual mode and tracked systems. Peter Peregrinus Ltd., Stevenage, UK.

FABRE, F, A KLOSE and G SOMER, Editors (1987). COST 302 Technical and economic considerations for the use of electric road vehicles. Directorate-General, Transport and Directorate-General, Science, Research and Development, Commission of the European Communities. Luxembourg: Office for Official Publications of the European Communities.

GYENES, L (1984). Advanced electric road vehicles: performance, range, energy use, and cost. Paper to University of Strathclyde (1984).

HARDING, G G (1977). Design of electric commercial vehicles for production. SAE Paper 770388. Society of Automotive Engineers Inc., Warrendale, Pennsylvania.

HARDING, G G and C MORRIS (1974). Silent Rider - a project for city centre transport. Paper in Proceedings of a symposium on battery electric road vehicles held at the Transport and Road Research Laboratory, 8 November 1973. Department of the Environment TRRL Supplementary Report SR 54 UC. Transport and Road Research Laboratory, Crowthorne.

HIRSCHENHOFER, J H (1989). International developments in fuel cells. Mechanical Engineering, August 1989, pages 78-83. American Society of Mechanical Engineers, New York.

HOUSE OF COMMONS (1989 I). Session 1988-89. Energy Committee Sixth Report. Energy Policy Implications of the Greenhouse Effect. Volume I. Report together with the Proceedings of the Committee. Paper 192-I. London: Her Majesty's Stationery Office.

HOUSE OF COMMONS (1989 II). Session 1988-89. Energy Committee Sixth Report. Energy Policy Implications of the Greenhouse Effect. Volume II. Memoranda of Evidence. Paper 192-II. London: Her Majesty's Stationery Office.

HOUSE OF COMMONS (1989 III). Session 1988-89. Energy Committee Sixth Report. Energy Policy Implications of the Greenhouse Effect. Volume III. Minutes of Evidence. Paper 192-III. London: Her Majesty's Stationery Office.

HOUSE OF LORDS (1980). Electric Vehicles. Report of the Science and Technology Committee, Session 1979-80. 1st Report. Paper HL 352. London: Her Majesty's Stationery Office.

HOUSE OF LORDS (1989 I). Session 1988-1989, 6th Report, Select Committee on Science and Technology. Greenhouse Effect. Volume I. Report. (HL Paper 88-I) London: Her Majesty's Stationery Office.

HUFF, J R, N E VANDERBORGH, J F ROACH and H S MURRAY (1987). Fuel cell propulsion systems for highway vehicles. Los Alamos National Laboratory, PO Box 1663, US Department of Energy, Los Alamos, New Mexico, USA. (Paper submitted to the 14th Energy Technology Conference, April 1987.)

JOEL, H F (1903). Electric automobiles. Minutes of Proceedings of the Institution of Civil Engineers. Vol. CLII. Published by the Institution, Westminster, London. (Reprinted by the Lead Development Association, London, WIX 6AJ.)

MITCHAM, A (1979). Hybrid electric vehicles - an assessment and survey. Paper 11 in Symposium on energy and road transport held at the Transport and Road Research Laboratory, Crowthorne, on 12-13 April 1978. Department of the Environment Department of Transport TRRL Supplementary Report SR 447. Transport and Road Research Laboratory, Crowthorne.

ROMANO, SAMUEL (1989). Fuel cells for transportation. Mechanical Engineering, August 1989, pages 74-77. American Society of Mechanical Engineers, New York.

SAUNDERS, D A (1976). Battery electric bus project. Final Report. Department of Industry, London.

SIMS, R I (1984). Advanced lead acid systems for LCEVS electric vehicles. Paper to University of Strathclyde (1984).
UNIVERSITY OF STRATHCLYDE (1984). Proceedings of the International Symposium: Advanced and Hybrid Vehicles, University of Strathclyde, 17-19 September 1984.
WATERS, M H L and I B LAKER (1982). The energy and driving patterns of a battery electric delivery vehicle in central London traffic. Proceedings of the International Congress "Drive Electric Amsterdam '82". AVRE Section, Netherlands.
WATERS, M H L and J PORTER (1974). A review of market prospects for battery electric road vehicles - part 1. Department of the Environment TRRL Laboratory Report LR 630. Transport and Road Research Laboratory, Crowthorne.
WATSON, R L, L GYENES and B D ARMSTRONG (1986). A refuelling infrastructure for an all-electric car fleet. Department of Transport TRRL Research Report RR 66. Transport and Road Research Laboratory, Crowthorne.
WEEKS, R (1978a). An investigation of a battery-trolley bus system. Department of the Environment Department of Transport TRRL Laboratory Report LR 823. Transport and Road Research Laboratory, Crowthorne.
WEEKS, R (1978b). A refuelling infrastructure for electric cars. Department of the Environment Department of Transport TRRL Laboratory Report LR 812. Transport and Road Research Laboratory, Crowthorne.
WICKEN, G W (1979). The London electric delivery van assessment scheme. SAE Paper 790111. Society of Automotive Engineers Inc., Warrendale, Pennsylvania.
WICKEN, G W and S MURRAY (1980). The London electric delivery van assessment scheme - some preliminary results. Paper presented to the "Drive Electric 80" Conference. Drive Electric 80, 30, Millbank, London, SW1P 4RD.
WOOD, R A, B R DOWNING, and T C PEARCE (1981). Energy consumption of an electric, a petrol and a diesel powered light goods vehicle in Central London traffic. Department of Environment Department of Transport TRRL Laboratory Report LR 1021. Transport and Road Research Laboratory, Crowthorne.
WOODWARD, BRIAN (1991). Academic engineers race to solar victory against Japan. New Scientist. Page 37, 12 January 1991. New Science Publications, Holborn Publishing Group, London.

Appendix A: Global warming and road transport

BOLIN, B (1986). How much CO2 will remain in the atmosphere ? In: Bolin, B *et al.* (Editors.) The greenhouse effect, climatic change and ecosystems. John Wiley and Sons, Chichester and New York.
BOLIN, B, B DOOS, JILL JAEGER and R WARRICK, Editors (1986). The greenhouse effect, climatic change and ecosystems. John Wiley and Sons, Chichester and New York.
BRITISH COAL (1989). British Coal Corporation Memorandum 3 to House of Commons (1989 II, page 15), and Evidence to House of Commons (1989 III, page 66). Also Memorandum to House of Lords (1989 II, page 156) and Evidence (page 158).
CCMC (1987). Impact of more stringent emission standards on vehicles with an engine displacement below 1.4 litres. CCMC Report AE/104/87, Committee of Common Market Automobile Constructors, Brussels.

C E G B (1989). Memorandum 5 by the Central Electricity Generating Board to House of Commons (1989 II, page 32).
DEPARTMENT OF ENERGY (1989). Memorandum 6 to House of Commons (1989 II, page 38)
DEPARTMENT OF THE ENVIRONMENT (1989a). Memorandum to House of Lords (1989 II, page 166)
DEPARTMENT OF THE ENVIRONMENT (1989b). Digest of Environmental Protection and Water Statistics, No 11 1988. London: Her Majesty's Stationery Office.
DEPARTMENT OF THE ENVIRONMENT (1989c). (In association with the Meteorological Office.) Global climate change. October 1989. London: Her Majesty's Stationery Office.
DEPARTMENT OF TRANSPORT (1989). Transport Statistics Great Britain 1978 - 1988. London: Her Majesty's Stationery Office.
EC (1983). Report of the ad hoc group ERGA - Air Pollution. Report III/602.83-EN-FINAL, Commission of the European Communities, Brussels.
GEORGE C MARSHALL INSTITUTE (1989). Evidence for natural cycles of warming and cooling, page 20, Scientific Perspectives on the Greenhouse Problem. George C Marshall Foundation, Lexington, Virginia, USA.
GIBBONS, J H, P D BLAIR and HOLLY L GWIN (1989). Strategies for Energy. Use In: Scientific American, Vol 261, No. 3, page 91, September 1989. Scientific American Inc., New York.
GRAEDEL, THOMAS E and PAUL J CRUTZEN (1989). The changing atmosphere. Scientific American, Volume 261, Number 3, Page 31, September 1989. Scientific American Inc., New York.
GREENPEACE (1989). Memorandum 9 to House of Commons (1989 II, page 62).
GRIBBEN, JOHN (1990). An assault on the climate concensus. New Scientist, page 26, 15 December 1990. New Science Publications, Holborn Publishing Group, London.
HANSEN, J, I FUNG, A LACIS, D RIND, S LEBDEFF, R RUEDY and G RUSSELL (1988). Global climate changes as forecast by Goddard Institute for Space Studies three-dimensional model. Journal of Geophysical Research, Vol.93, No.D8, pages 9341-9364, August 20 1988. American Geophysical Union, Washington, DC.
HOUGHTON, RICHARD A and GEORGE M WOODWELL (1989). Global climatic change. Scientific American, Vol.260, No.4, pages 18-26, April 1989. Scientific American Inc., New York.
HOUSE OF COMMONS (1989 I). Session 1988-89. Energy Committee Sixth Report. Energy Policy Implications of the Greenhouse Effect. Volume I. Report together with the Proceedings of the Committee. Paper 192-I. London: Her Majesty's Stationery Office.
HOUSE OF COMMONS (1989 II). Session 1988-89. Energy Committee Sixth Report. Energy Policy Implications of the Greenhouse Effect. Volume II. Memoranda of Evidence. Paper 192-II. London: Her Majesty's Stationery Office.
HOUSE OF COMMONS (1989 III). Session 1988-89. Energy Committee Sixth Report. Energy Policy Implications of the Greenhouse Effect. Volume III. Minutes of Evidence. Paper 192-III. London: Her Majesty's Stationery Office.
HOUSE OF LORDS (1989 I). Session 1988-1989, 6th Report, Select Committee on Science and Technology. Greenhouse Effect. Volume I. Report. (HL Paper 88-I) London: Her Majesty's Stationery Office.
HOUSE OF LORDS (1989 II). Session 1988-1989, 6th Report, Select Committee on Science and Technology. Greenhouse Effect. Volume II. Evidence. (HL Paper 88-II) London: Her Majesty's Stationery Office.

MATTHEWS, ROBERT (1989). New risk from CO2 revealed. The Sunday Correspondent, page 11, February 4 1990. The Sunday Correspondent Limited, London.
METEOROLOGICAL OFFICE (1989). Memorandum to House of Lords (1989 II, page 57 and 58)
NERC (1989). Natural Environment Research Council Memorandum 11 to House of Commons (1989 II, page 76).
RAMANATHAN, V, R J CICERONE, H B SINGH and J T KIEHL (1985). Trace gas trends and their potential role in climate change. Journal of Geophysical Research, Vol.90, No.D3,pages 5547-5566. American Geophysical Union, Washington, DC.
RAMANATHAN, V (1988). The Greenhouse Theory of Climatic Change: a Test by an Inadvertent Global Experiment. Science, Vol.240, pages 293-299, April 1988. American Association for the Advancement of Science, Washington D C.
"THIS COMMON INHERITANCE" (1990). This Common Inheritance. Britain's Environmental Strategy. Presented to Parliament by the Secretaries of State for Environment, Trade and Industry, Health, Education and Science, Scotland, Transport, Energy and Northern Ireland, the Ministries of Agriculture, Fisheries and Food and the Secretaries of State for Employment and Wales. Cm 1200. September 1990. London: Her Majesty's Stationery Office.
THRUSH, B A (1989). Memorandum to House of Lords (1989 II, page 271)
UNIVERSITY OF EAST ANGLIA (1989). Evidence to House of Lords (1989 II, pages 1-38)
WARREN SPRING LABORATORY (1989). Memorandum to House of Lords (1989 II, page 282)
WARRICK, R A, E M BARROW, and T M L WIGLEY (1990). The Greenhouse effect and its implications for the European Community. Report EUR 12707 EN, Directorate-General, Science, Research and Development, Commission of the European Communities, Luxembourg.
WORLD WIDE FUND FOR NATURE (1989). Memorandum 17 to House of Commons (1989 II, page 109)

Appendix B: Energy considerations for other modes of transport

ACEC (1976). Passenger transport: short and medium term considerations. Advisory Council on Energy Conservation Paper 2. Energy Paper Number 10. Department of Energy. London: Her Majesty's Stationery Office.
HAMMARSTROM, ULF (1988). Equivalent energy consumption and exhaust emissions. Swedish Road and Traffic Research Institute (VTI) Paper 567. Linkoping, Sweden (in Swedish, with English summary.)
HILLMAN, MAYER and ANNE WHALLEY (1983). Energy and personal travel: obstacles to conservation. Report No.611. Policy Studies Institute. London.
HIRST, ERIC (1973). Energy intensiveness of transportation. Transportation Engineering Journal, Proceedings of the American Society of Civil Engineers. February 1973, pages 111-122. Washington, DC.
ISGLUTI (1988). Urban land-use and transport interactions. Policies and models. Report of the International Study Group on Land-use/Transport Interaction (ISGLUTI). Edited by F V Webster, P H Bly and N J Paulley. Avebury, Gower Publishing Co. Ltd., Aldershot, Hampshire, and Vermont, USA.

LANGDON, M G (1982). Multiple choice models in transport assessment. Department of the Environment Department of Transport TRRL Laboratory Report LR 1048. Transport and Road Research Laboratory, Crowthorne.

LEACH, G, C LEWIS, F ROMIG, A van BUREN and G FOLEY (1979). A low energy strategy for the United Kingdom. The International Institute for the Environment and Development. Science Reviews Limited. London.

MALTBY, D, I G MONTEATH and K A LAWLER (1978). The UK surface passenger transport sector. Energy consumption and policy options for conservation. Energy Policy, December 1978, pages 294-313. Butterworth-Heinemann, Oxford, UK.

MARTIN, D J and R A W SHOCK (1989). Energy use and energy efficiency in UK transport up to the year 2010. Report for the Department of Energy by the Chief Scientist's Group, Energy Technology Support Group, Harwell.

"THIS COMMON INHERITANCE" (1990). This Common Inheritance. Britain's Environmental Strategy. Presented to Parliament by the Secretaries of State for Environment, Trade and Industry, Health, Education and Science, Scotland, Transport, Energy and Northern Ireland, the Ministries of Agriculture, Fisheries and Food and the Secretaries of State for Employment and Wales. Cm 1200. September 1990. London: Her Majesty's Stationery Office.

WPICT (1970). Comparative assessment of new forms of transport. A Report by the Interdepartmental Working Party on Inter-City Transport (Department of the Environment, Department of Trade and Industry, Ministry of Aviation Supply). Reissued in April 1973 as TRRL Supplementary Reports SR 1, 2, and 3. Transport and Road Research Laboratory, Crowthorne.

Subject index

Printed in the United Kingdom from HMSO
Dd.295036, 3/92, C10, 3390/3, 5673, 186049